a curious guide to
the natural history of the
pehuén

Jane C. Coffey

three degrees of latitude

a curious guide to
the natural history of the
pehuén

Photos by Jane C. Coffey
Designed by Anna Pillow
Printed in the United States of America

Philosophically considered, the Universe
is composed of Nature and the Soul.

●

Ralph Waldo Emerson

perspective

In the quiet of an approaching fall day perspective can shift rapidly at first light, particularly on a fall day in March. In the blue-blackness of twilight the edges of things start showing up; silhouettes start appearing. A flat, dark expanse slowly gains the depth it needs to be another day. Dawn can be the softest of moments as the earth revolves us out of a still night and gently into the sun. It is a thoughtful space of time for the universe to give us, but sometimes our primal response to this perfect quiet is to question our presence on this earth, our place in this deep universe.

It was in a similar still moment, toward the end of a previous and more northern twilight, that I once bothered to think about something as big as the big bang theory. I understood, as dawn quietly sneaked in that other morning, that this theory sounded too loud, too startling an occurrence for the beginning of any new day. Now here in the darkness of another hemisphere I crouch on a silent rocky hillside in the Andes Mountains looking out across the floor of an ancient and collapsed caldera and my thoughts leap again from this quiet. This time, I leave behind the beginning of the universe, I go beyond the creation of earth, through 4,000 million years of the Cryptozoic Eon, past another 325 million into the era of the Paleozoic to stop, stunned, looking out across this valley into what appears to be a chunk of time that began 200 million years or so just shy of the present. This dawn looks to be the beginning of just another day in the Jurassic.

In this cool Argentine morning I watch as the rim of the caldera, now a ring of high and weathered blue mountains, slowly separates from the sky. Close to where I sit, water cascades over worn chunks of basalt and flows away from me down into the valley. I already know about this river and its acid nature and watch as it spreads across the valley floor and flows into a lake. Its source, an active volcano, sits behind me quiet for the moment and crested with snow. With the sky reflecting metallic in the two fingers of the lake and in each meander of the river, I watch the fog lift from the valley revealing layer by layer the raw, jagged outlines of volcanic rock. There is, this morning light shows me, not one green thing that edges in from the surrounding earth to be nourished by these waters.

Green does exist in this valley, but not near this river nor down by that lake. Out of the sienna, gray, or dusty-black igneous rock trees do grow. Most species found here, at this altitude, in this soil, never grow to be more tree than bush. Here in this place only one species can claim the height of a tree but that tree, *Araucaria araucana*, certainly grows tall. Here on this morning, watching the sun slide in behind these odd, ancient conifers, my perspective on tall is altered, my sense of time displaced. This morning, just as on some other morning maybe 80 million years ago, it seems quite possible that a Patagosaurus could round this outcropping of columnar basalt, pound and thump its way past beneath the high, umbrella-like crowns of these trees and splash on through that flow of acidic water. Although deeply absorbed in this shift of perspective, this visual displacement of time, I do understand that the ancestors of these grand trees around me outlived these huge creatures.

I do know it is a fall morning in the late Cenozoic, but still I sit and wait a few hours — nothing thunders past.

If, in crossing the Andes, Charles Darwin had taken this particular mountain pass from Argentina into Chile instead of a short ways farther north through Mendoza, I am afraid his theory of evolution might never have truly evolved. It would have been a difficult conclusion to come to, studying these trees and this valley. Darwin writes in his journal about finding an ancient stand of petrified *Araucaria* on the slopes of the Uspallata range in Argentina. If after, on his way into Chile, Darwin had passed through Copahue with its forest of *Araucaria* and its pre-historic view he certainly might have made another entry, a remark or two, about time *without* change. As it is now, I can only sit here and imagine him happening upon this valley, exploring this caldera, misdirected here by some random shove of fate. As for me, I have not come by chance to sit on this hillside. I did not happen upon Caviahue. This town, sitting below the Volcán Copahue on the boundary of Argentina and Chile, would be a difficult place to stumble across. I have come here to see the trees, the pehuéns, *Araucaria araucana*.

seeing

Seeing is a phenomenon different than sight. A painter looking out at a landscape sees a dot of red, maybe a rock, vibrating in a swath of green. A geologist standing in the same spot may not notice this juxtaposition of color but does see the forces of time in that rock. Ralph Waldo Emerson understood this difference in seeing; he explained it in this manner — *we animate what we can, and we only see what we animate*. (Emerson, 1990) When arriving in any new place what I isolate and pull from the landscape first are the trees. It is a simple matter of attraction. Some people are first drawn by their senses to what is different about a place: colors, smells, temperature. Others look at who or what might inhabit a spot: people, animals, structures, existing in their wild, urban or suburban forms. Still others notice first what haunts a place: history, beauty, poverty, wealth or the blaring commerce of tourism — what I am first attracted to, for some reason, are the trees. As a very young child I saw trees as wonderful spots to hide and as an adolescent they seemed perfect places to sulk. Now, as an adult, when I am not hiding or sulking in them, I study them. Usually, I wait until a tree catches my eye to investigate its nature. I might notice an outline, unique in its own graceful, grand or sometimes gloomy way. Or how a sycamore with its peeling, splotchy trunk smoothes and stretches upward into the taut slick green of a new branch. At times it is a photograph or an illustration that catches my eye, at other times a story, or a description sends me looking. In fifth grade I demanded to be the spinney of larch trees in the drama of Pooh's Corner, soon after to run home to find out what a larch looked like and what exactly was a spinney. That word "big" once drove me up the coast of California to stand under a redwood — a friend's emphatic use of this word was not an exaggeration and my image of *big* grew taller.

But to lean against the rough bark of this big southern conifer, to see the thick triangular needles of the pehuén, I had to fly to the Southern Hemisphere, traverse the beautiful and broad city of Buenos Aires, fly to the northwestern edge of Patagonia and take a dusty, six-hour bus ride up into the volcanic landscape of Copahue. A long journey started from a short description, a brief, simple description, but a description, I must say, well delivered.

Jelle Zeilinga de Boer and his colleague, Johan Varekamp, have both seen much of the world; they have many stories to tell, they get excited about a lot of things. As geologists, more specifically volcanologists, they are quite busy out in their field. They have a research project in northern Patagonia that revolves around the small but active Volcán Copahue. In sharing their stories of this place with me their excitement always rose and emphasis always fell on the word *unusual*. In this spot, a rather young 10,000 year-old volcano sits in a dramatic landscape formed by a caldera that collapsed some two million years ago. A glacier folds over the west rim of the crater and hangs above a lake that steams and roils and belches clouds of hydrochloric acid. A river, the aptly named Rio Agrio, runs acidic water down from the volcano, through the valley and into another lake, a calmer lake. This lake, scoured out of the caldera floor by glaciation is Lago Caviahue. It is said that Caviahue is an indigenous word to indicate a meeting place, but sitting on the shore of this body of water there is not much to meet—nothing swims, squirms or even crawls slowly by.

The two geologists had marvelous stories about their travels to this remote spot but throughout all their dramatic descriptions, it was this tree I kept trying to imagine, it was this tree I knew I wanted to see. An unusual tree, of course, a tree that can grow 165 feet tall straight out of volcanic rock, a tree that can live for 1300 years, a tree whose direct lineage is counted back in millions of years, a tree that has not changed much since the Jurassic Period. An ancient and towering conifer that has thick, triangular, scale-like leaves, deeply textured bark and pine nuts two inches long encased in a cone the size of a child's head. In Caviahue they call this tree the pehuén, pronouncing it the same as the indigenous spelling, pewen. A Jesuit botanist from Chile living in exile in Italy gave it the scientific name *araucana*. The English, in the 1800s, nicknamed it the monkey-puzzle tree. The locals sometimes called el paraque, the umbrella tree—it is a species with many names. The professors were planning to go back to Caviahue and invited me to tag along, take photos and possibly do research for a thesis. But then, unfortunately, their trip was canceled. It all happened rather quickly, once I decided to go alone. I promised to bring back the water samples they needed—so they handed me maps to consult and diagrams to follow, a name here, a contact there and a large metal suitcase full of water sampling devices. The gas mask, unfortunately, they could not find.

A GENERAL MAP OF THE
WITH ALL THE NEW DISCOVERIES and MARGINAL DELINEATIONS, CONTAINING TH
Published by LAURIE

WESTERN HEMISPHERE OR THE NEW WORLD with the New Discoveries and Several Additions

The NORTHERN HEMISPHERE

NORTHERN ICY OCEAN

NORTH AMERICA

ATLANTIC or WESTERN OCEAN

NORTH PACIFIC OCEAN

PACIFIC OCEAN or GREAT SOUTH SEA

GULF OF MEXICO

SOUTH AMERICA

SOUTH PACIFIC OCEAN

SOUTHERN ICY OCEAN

AN UNIVERSAL SCALE

A CHART of the WORLD according to MERCATOR's PROJECTION With the Rhumb Lines

BAFFIN BAY

FROZEN SEA

Siberia

Russian Tartary

PACIFIC AMERICA

NORTH SEA

proportion

When hiking in any mountain range a sense of proportion is an important thing. It is useful to judge time, distance and scale by visually comparing the known to the unknown—a human figure far ahead on the trail, that roof in the valley below, the size of the trees on the next ridge. How big are they, how small, how far is left to go? Judgments can certainly be made by assessing proportion and determining scale but standing out in the three dimensions of this spherical world, proportion can be a tricky thing. When placing one's self on the edge of the Grand Canyon, its vastness might seem apparent but it is not until a person descends that long, thin, dusty trail down the steep canyon wall to stand on the bank of the Colorado River looking up can one truly understand the scale, the depth of this grand cleft in the earth. Depth, or height, cannot be misunderstood from a point of view such as this. With an ancient redwood, or here in these Andes, a pehuén, it is the same.

Experiencing the true proportions of nature can be important in acquiring an accurate sense of scale but conveying that information visually is often not an easy thing to do. A painter might create a landscape of Caviahue depicting the proper scale of volcano to valley, valley to lake, giant trees to small village but anyone viewing this work might see the artist's attempt at conveying the tallness of the trees in this scene as faulty proportion—something not quite right. They might not put trust in a two dimensional surface. It is known that photographs can lie about proportion but even given proper visual clues, awe might not be inspired. As impressed as one might be by a photograph of a car driving through an arch cut in the trunk of a redwood, it is novelty that is being illustrated. The physical impact of greatness or grandeur, that stab of nature's imposing presence, is not well illustrated. Such concepts might do better in the hands of an artist — an artist who disregards proportion altogether.

To a European artist living in the 13th and 14th centuries the perspective of the human eye did not hold much importance. Proportion was a relative thing, relative to, not the physical world but to a social-spiritual hierarchy. There was no guessing at that time what a painter and his patron, the Catholic Church, saw as important — divinity was drawn large. In a worldview such as this an artist such as Cimabue, one of the few identified Gothic painters, portrayed the holiest of his subjects the largest, then scaled on down to any mere mortal; nature, along with ordinary human beings, buildings, maybe even volcanoes, were small things. Here in Caviahue, the reality of these tall trees might be better conveyed in two dimensions if one took proportion in hand — there would be no question on the grand scale of these trees, no historical room for doubt, if Cimabue had been a Mapuche instead of a Florentine.

Here the Mapuche revere this tree, it has always been a large presence in their lives. The spiritual tradition of those Europeans might have deemed nature's presence a small thing in their view of the world but for the people who are indigenous to these mountains this tree has always loomed large in their lives. Pine nuts were once necessary to their survival; even still some of the bread they make, the stew they cook and a fermented beverage they drink, comes from these seeds. There are Mapuche who live in Chile and Argentina in the regions where these trees grow who refer to themselves as Pewenches, people of the pewen. This bond is not a small one. The Pewenches (Pehuenches) still collect these pine nuts though they no longer subsist on them but the tree is part of their culture and theirs is a culture rooted to the land. Now, in a narrow swath of Chile and Argentina, between 40° 20 and 37° 20 latitude there is some effort to keep the isolated patches of this species alive. The pehuén, in just these three degrees of latitude, continues to hold its place, though its future is hard to predict. The Mapuche, unfortunately, are left to struggle for their cultural survival all on their own.

Surviving the Spanish invasion was no small feat but the Mapuche had gained much practice battling the Incas. The Incas waged many wars against the Mapuche trying to expand their territory into what came to be apportioned by the colonizers as Chile and Argentina but after many brutal defeats the Incas decided it might be best to keep their distance. It can be said that the European colonizers were also not able to defeat the Mapuche, but keeping their distance was never a consideration. There were frequent battles and many deaths and the Europeans continued to encroach, taking more and more land.

The colonizers found the Andes much harder to access then the coastal areas and in part, the survival of the pehuén seems to have been aided by its presence in a difficult terrain; what was hard to reach was more likely to be left alone. The Mapuche, however, were not so lucky, though their language and culture survive, a rugged terrain did not prevent them from losing much of their territory, even in these steep Andes Mountains.

Caviahue is close to the northern-most range of the pehuén in Argentina and the area around Caviahue is one of the parks created by the province of Neuquén to protect what is left of this species in Argentina. At the time, Ernesto Maletti was the agriculture engineer for the *Area Natural Protegida Parque Provincial Copahue/Caviahue* and I was told he would certainly know about the pehuén. I went looking for him the morning after my arrival in Caviahue. Small, isolated towns like Caviahue require little footwork to track a person down. Asking a random individual how to find a local resident will almost always result in information. It could be just a vague flick of the hand to indicate the way down the road or maybe more precisely, a drawing etched in the dirt or maybe on some cool, crisp morning a little local history might be provided along with an escort to that person's door. Ernesto was sitting in an office when I found him, but his dusty boots, jeans and brown fedora announced immediately that this was *not* where he spent his time.

This slightly bearded, young man listened intently to the rambling explanation of my presence in Caviahue. He nodded several times but he did not say much except *bueno*. He repeated this several times to himself or to me, I wasn't quite sure, as he wandered around the office looking for something, lifting piles of papers, opening draws and closets. Finally, he stopped abruptly, headed for the door and asked if I'd like some lunch. Five minutes after meeting this man I sat in his kitchen. His young wife, Analia, showed little surprised at my appearance at their table and as a stranger I sat very comfortably in her house. In the weeks to come I spent many hours at that kitchen table, learned many things about their lives. Analia, just finishing her thesis, will be a civil engineer. She is warm and energetic and is not thrilled about the isolation of Caviahue. Ernesto is interested in the world; he asks many questions, writes poetry and is a singer of folkloric songs. His religion is very important to him. His order of the spiritual world, Catholicism, is in the same tradition as Cimabue's. His esteem of the pehuén comes close to that of the Mapuche and what I soon understood was that, for him, no matter whether seen in two or three dimensions, figuratively or literally, these trees stand very large.

Caviahue is a small place overshadowed not by the presence of Copahue the volcano as one might expect but by Copahue the spa, or "Las Termas." Most people come to the Copahue/Caviahue area to sit in the thermal waters that bubble to the surface near the volcano. Once, it effervesced as a natural spa under the stars and the domain of the local Mapuche tribe. Now, these waters are sheltered by cement buildings and controlled by the Argentine government. Doctors walk the halls. People come from all over South America and as far as Europe to soak away various ailments. There are no trees of any kind in Copahue, just gray, volcanic soil and steaming, bubbling, pools of smelly water. Ernesto knew I was not a tourist seeking out the curative waters here; very few North Americans show up in this town without some interest in the volcanics of this place. Though I would be taking water samples from the lake, river and the volcano's crater lake I was not a hydrologist. Through the jumbled syntax of my Spanish I tried to explain why I had really come but I wasn't sure he quite understood. The following day, it became apparent he had, perfectly well.

That next morning, Ernesto met me with maps and charts in hand. *Parque Provincial Caviahue/Copahue Area Natural Protegida* he laid out before me contours and all. Ernesto pointed out the boundaries and the areas shaded in different colors on the map. This was a map of his own making, something quite evident by the way he carefully pointed out its details. It was a map that illustrated what was important to Ernesto about this land. Color-coded squares and rectangles delineated the zones of high use, those of protected use and a small portion of land that was off limits to all. Here, he told me, his finger tapping on this small spot of the map, he is reforesting with *Araucaria araucana*. As he spoke, his eyes, fixed and luminous, watched mine.

I recognized the moment; I understood the look. It is that moment when someone presents to you their passion and then pauses a flicker of a second looking for a sign of recognition, a semblance of understanding. No matter whether one is a scientist, a naturalist, a poet or a painter, there is nothing better than a moment like this, when recognition is found and sensibilities shared. Caviahue, at that moment, did not seem so far from home, or home, perhaps, did not feel like such a singular place. I realized standing in the dry heat and red dust of this place I had found a kindred spirit, another moment in time and another spot on earth to link myself to.

scale

Scale translates one image for another or allows us to figure out the real world made flat. But the distance from one hemisphere to the other, regardless of the scale on the bottom of the map, is a relative thing. I learned about maps and geography in a tiny classroom in a small town that most mapmakers never even gave a dot for. In my young mind this huge chunk of important, but missing, information, my entire world in fact, created in me a particular mistrust of maps. I can spend hours looking at these drawings with their lines and legends, colors and ratios, their intimation of time and distance, but I always view them as abstractions and not fact. Once they have proven to lead me to where it is I want to go, I am more trusting. But in trying to reach Copahue I could find not a line, solid or dashed that would take me up into the area of the Andes where I needed to go. In fact, the maps showed me very little until I got as far as the city of Neuquén, near the foot of those mountains. Everyone there knew where Copahue was and which line to take to get there. There was only one. But my journey had started back home, without even this one line to follow and no dot to head for. The maps I had perused back home showed only a triangle, a solid triangle—the cartographer's symbol for an active volcano. I have a particular fondness for getting to know dotless places, especially dotless places with unusual trees, but solid triangles—they can be a very different matter.

Straddling the western most edge of the province of Neuquén there are quite a few of these triangles. In fact, in the Andean Cordillera from the Columbia/Venezuela border all the way down to La Isla Grande de Tierra del Fuego the maps are full of them. Chile and Argentina share a very long border that snakes around and over the tops of these mountains and near many of these volcanoes. Copahue is just one of them. Here, the border of Argentina juts into Chile and passes around the west flank of this volcano, logically situating the source of a possible eruption in the same country as the town it might someday wipe out. The town of Caviahue sits in Argentina, below Volcán Copahue and its crater full of hot water. Caviahue has only been a town since 1986 and did not exist when the national boundaries were being decided.

The first treaty to establish borders was drawn in 1881. Though there has never been a fight over this crater lake, borders, like fences, exist to argue over and Chile and Argentina have only recently finished bickering over their mountainous backyard. It was not until December 1998 that the last border of unclear definition was set across what is known as Heilo Sur, a massive, but shrinking, sheet of glacial ice in southern Patagonia, ending at least one hundred years of squabbling. This particular dispute may be over but the two countries are anything but pals. The Argentines are still sore that the British military was allowed to land in Chile — to land in Chile on the way to fight the Argentines over Las Islas Malvinas as they are called in Argentina. The British like to call them the Falklands, they also still very much like to call them their own.

The border back in 1881 was defined as following the highest peaks that divide the waters. *Divortium aquarium* — which way the water shed — became the guideline to who owned what. Of course, nature very seldom accommodates politics and as it turned out the highest peaks in the southern Andes do not always form the watershed. Glaciation during the Pleistocene Age left terminal moraines east of the mountains that diverted eastward flowing rivers back toward the Pacific — not so good for Argentina. In 1893, borders still in dispute were drawn across the summits of the highest mountains. This is after Argentina, in one case, not about to lose the large area around what is now Lago Buenos Aires in the province of Santa Cruz, cut through an ancient terminal moraine so water would run out the east side of the lake to the Atlantic. Argentina then could claim half of the lake and the surrounding huge tract of land as theirs — *divortium aquarium* man-made.

But in Caviahue the borders are clear, the Argentines have the thermal waters of Copahue all for themselves. The border curves around the base of the volcano and if the crater lake overflowed it would do so down the Argentine side. A gash in the side of the volcano also happens to be on the southeastern flank of the mountain. A hot, yellow stream of acid brine boils out from a fracture in the crater wall and steams down the side of the volcano. It widens into the Rio Agrio and starts to make its inevitable journey toward the Atlantic Ocean. *Divortium aquarium,* the direction of flow of this river is why Copahue is in Argentina, why two professors from Wesleyan University are doing research in Copahue and why I wound up lugging around a suitcase full of water sampling equipment to South America. Highly acidic water, pH1, leaves Volcán Copahue but it is fresh water that shows up at the Atlantic. In studying the hydrothermal system associated with this volcano and its drainage, Professors de Boer and Varekamp are trying to figure out the roles dilution and/or rock water interaction play in the neutralization of this acid water. They are also keeping an eye on the water in this crater. Active volcanoes themselves need keeping an eye on but a crater lake throws in another worry.

Crater lakes by nature are unstable things. What is necessary for their existence, besides an appropriate contour to hold water, is a structure that has walls that can effectively contain it as well. Stratovolcanoes like Copahue are better at this than some, their penchant for spitting ash creates matter that can find its way into fissures and fractures and lessen any seepage from a crater filled with water. Lava-flowing, basaltic volcanoes like those that are the Hawaiian Islands are good at creating new real estate but the lava is porous and not so hot at containing lakes. Once the form is there, however, content is still a problem. Any kind of balance between heat flux from the volcanics below the lake and atmospheric cooling is not a common state. The odds of the rate of evaporation being offset by precipitation makes staying a body of water in the crater of a volcano a precarious thing. Volcanoes on the whole are precarious things; living below a crater lake ups the ante a notch or two.

I have long since given up trying to figure out why people choose to live on fault lines, below volcanoes, or in trailers parks in spots known as "tornado alley." A sense of fatalism coupled with a pronounced gambling nature, perhaps, gives rise to that particular philosophy of what happens, happens — but chances are it won't happen to me. Waking up in the twenty-first century every morning contains enough of this experience for me not to compound it by living in any sort of natural disaster zone, but then, I grew up on the relatively stable bedrock of New England near the usually calm shores of Long Island Sound. I find tectonic and climatic forces have strong, attractive, often hypnotic powers but I have never been subject to them at their worst.

Isolated catastrophic incidents can change local environments, as when a category five hurricane denudes an island of most of its flora and fauna or a volcano erupts and buries all beneath its molten flow. Earthquakes, tsunamis, tornadoes, are all horrific and devastating events, events that change or end individual lives, events that happen quickly, but geological and climatic events that change life on earth, end species, take time. It is a length of time that the human mind stumbles on. As relative newcomers to this earth, our point of view of geological and climatic events in our environment is skewed. We have no adequate scale.

point of view

The last two hundred years, a short time even on the human scale of things, has colored how we view extinction. In this alarmingly short period, humans have brought about the end of many species and pushed countless others to the brink of extinction. Teetering on this edge of survival, as many are, sudden natural events might in the end push them off, but these forces cannot be said to have caused the extinction. Because we have been so busy altering our environment so drastically these last few hundred years, a natural extinction as opposed to one caused by the human factor is something we know little about.

If the last several hundred years have shaped our view of extinction, the last several decades have grounded our

point of view. The study of biogeography, particularly island biogeography, has preoccupied us since we woke up to the fact that we were altering the environment to such a degree that although, for the most part, we have actively ceased killing off whole species, we were continuing the extinction by wiping out their habitats. Much debate exists in study after study of population and extinction; was it better to set aside a large number of small "natural" areas or more effective to have fewer but larger chunks of habitat? What species are most at risk? Which are more likely to survive? Island ecology became the proving ground for many studies, but perhaps the study of island biogeography with its many examples of relatively rapid adaptation through evolution has colored how we look at the survival of any species.

As members of a species that spends much time digging around for our long-limbed, heavy-lobed predecessors, we are particularly preoccupied in a theory that labels the time between them and us as an evolution. Charles Darwin and Alfred Wallace both spent much time thinking about why and how things change, how they evolve. They both gnawed away at this subject until they came up with a process known as natural selection — the survival of the fittest. We like the idea of eons, eras and epochs of the fittest genes being filtered down to become us. Even those folks whose theology conflicts with where Darwin's idea of the origin of our species leads back to have no problem embracing this idea of survival of the fittest. But does our highly evolved species point of view influence how we look at the survival of other species, particularly other species that did not change or evolve into something else?

A species, like *Araucaria araucana*, left over from a time when conditions on earth were very different is considered a relic species. Archaic is another term used; *belonging to former or ancient times* — this is not a pejorative definition but in the sense of not being highly evolved it is often used as such. Sometimes we are so preoccupied with the razzle-dazzle of change we forget to notice the things that have stayed the same. We forget to appreciate the fact that they have survived staying the same. We forget to wonder *why* they might have stayed the same.

The pehuén is only one species in a whole family of archaic trees. This family of southern hemisphere conifers, *Araucariaceae*, consists of two genera: *Agathis* and *Araucaria*. In the genus *Araucaria* there are nineteen species found in New Caledonia, Northern Australia, New Guinea, the Pacific Ocean Islands and South America. Of these nineteen only two are South American species, *Araucaria angustifolia*, found in the sub-tropic and temperate rainforests in southeastern Brazil, Paraguay and northeastern Argentina and *Araucaria araucana* found in the temperate latitudes of Chile and Argentina. Fossils from this family, however, have been found throughout the northern hemisphere — these trees once spanned the globe. Other gymnosperms such as the gingko and the cycads have also survived for millions of years changing very little; the pehuén is not alone. There are other species besides these trees that have not changed much, not evolved, that are closely tied to the pehuén — a hand full of beetles for instance. These beetles, in the genus *Mecomacer*, are currently found under the thick bark of the *Araucaria araucana*. They consider it home and have been doing so for a very long time. They too are of an ancient lineage.

Fossil *Araucaria* have been found in Argentina munched on by the ancestors of these same species of *Mecomacer*. Insects, unlike trees, are known to evolve quickly, the 330,000 species of beetle in the world today certainly underscore that point, but the beetles associated with *Araucaria araucana* did not change. This is one of the oldest plant-insect antagonism known. (Farrell, 1998) Both these species have survived remaining relatively unchanged. What allows for this seemingly stable relationship, a relationship of two million years or so, is not yet clear but it underscores the sense of balance this species of tree has maintained over the course of a changing global environment.

There is much fascination for life that has disappeared from earth apparently due to natural forces. Dinosaurs and their fate continue to intrigue us. But how often do we think of these dramatic forces in terms of the species that survived them? Why are we here driving snowmobiles around the Arctic Circle without a Mastodon in sight? Or why are the *Araucaria* here but no dinosaurs grazing underneath? Sitting on volcanic rock, in a grove of towering pehuéns, where a deep patch of soil is a difficult thing to find, this all might cross one's mind. When this grove is on a slope of an active volcano, below a glacier, streaked gray with ash, thinking of this spot as a place where an endangered species could, and does, survive seems a wondrous thing.

Darwin did not cross the Andes back into Chile through Caviahue; he took the Uspallata Pass further north in Mendoza, the pass just south of Aconcagua, the tallest mountain in the Western Hemisphere (22,831 feet). No *Araucaria* grow in that area now but while trekking through the Uspallata Range in Chile Darwin came across a stand of petrified trees.

> ....at an elevation of about seven thousand feet, I observed on a bare slope some snow-white projecting columns. These were petrified trees, eleven being silicified, and from thirty to forty converted into coarsely-crystallized white calcareous spar. They were abruptly broken off, the upright stumps projecting a few feet above the ground. The trunks measured from three to five feet each in circumference. They stood a little way apart from each other, but the whole formed one group. Mr. Robert Brown has been kind enough to examine the wood: he says it belongs to the fir tribe, partaking of the characteristics of the Araucarian family. (Darwin, 1969 ed.)

Darwin makes no comment on the relationship of these fossilized trees to any living *Araucaria araucana* he might have seen in his travels through Argentina and Chile; it is possible he never saw a pehuén. Darwin's point of reference in these mountains, his point of view, was seemingly influenced by what he was reading at the time. He sees these petrified tree stumps and goes into great detail on his perception of the geology of the Andes Mountains, surmising how these trees, *once waved their branches on the shore of the Atlantic, when the ocean (now driven back 700 miles) came to the foot of the Andes.* (Darwin, 1969 ed.)

On board the *Beagle* Darwin had brought with him to read the recently published first volume of Sir Charles Lyell's *Principles of Geology* and the Andes Mountains were certainly providing Darwin the perfect backdrop for reading Lyell's work. In this work, Lyell expounds at some length on the forces that he saw as originally shaping the earth (erosion and sediment accumulation, earthquakes and volcanoes) and how he sees these as being the same forces that were currently at work. The second volume of Lyell's work was waiting for Darwin when his boat docked in Montevideo, Uruguay. A third volume arrived in Darwin's hands on the other side of South America in Valparaiso, Chile, hot off the press. His Cambridge botany professor, John Stevens Henslow, had sent them along with the caveat not to believe all that they contained. Here Lyell expands his ideas on the uniform nature of geology and transfers them into the realm of biology. To Lyell this meant there would always be gradual but uniform change on this earth and no need to imagine the catastrophic to explain these changes. The ongoing discovery of fossils and the extinct species they often represented were giving rise to many theories as to the disappearance of these species and Darwin's fascination is clear. *Certainly, no fact in the long history of the world is so startling as the wide and repeated extermination of its inhabitants.* (Darwin, 1969 ed.)

At that time, evolution did not play a part in Lyell's view of a species continuing through geological time. He saw species as stable entities that could be driven to extinction but not altered. He would come to change his tune years later after reading the work of both Darwin and Wallace and their views on the subject but at the time he felt that the survival of a species depended on a particular set of physical conditions and geological processes were constantly changing those conditions. Hence, failure to compete with other species in a new habitat might extinguish a species and the success of one prosperous species (wisely Lyell included *Homo sapiens* in this equation) might crowd out others to extinction. In *Principles of Geology*, in summing up his ideas on extinction at that time, Lyell writes:

From what has now been said of the effect of changes which are always going on in the condition of the habitable surface of the globe, and the manner in which some species are constantly extending their range at the expense of others, it may be deduced, as a corollary, that the species existing at any particular period, may, in the course of ages, become extinct one after the other. 'They must die out' to borrow an emphatic expression from Buffon, 'because Time fights against them.' (Lyell, 1877)

If Darwin had come across Caviahue with its volcanic landscape and its striking example of a monospecific pehuén forest, would his point of view at that time have caused him to perceive these trees as "fit" competitors in this environment or would he have considered these trees primitive entities whose time was running out?

observation

When learning things through observation, whether it is in reference to nature or to art, the relationship between perception and subject matter is always in a state of flux, should always be in a state of flux. The more we look, the more we learn; the more we learn, the more we see. In art, perception fuels the creative process only if the act of perceiving does not become a fixed moment. When examining nature, the same is true. If our observations come with expectations or limits then we lose a most important tool of learning — we lose that pure moment of observation.

As a photographer I have long been aware of the stimulation of the unknown environment. When seeing things for the first time, one's visual sense is not interrupted or overruled by information. With the unfamiliar, the visual elements are more accessible; details are more likely to catch one's attention. A uniquely shaped tree or the interesting shadow that it might cast can share equal billing. Take that same tree and place it on a street you walk down every day, its form, the visual elements of line, shape, and pattern, might disappear into that generic clump of a word *tree*. This phenomenon makes taking an interesting photograph in one's own backyard, even when there is a grand spectacle in that backyard, a great feat. It means that in hours, days, years, of looking at the same view, *seeing* can still take place — it is not superseded by *knowing*.

My first few days in Caviahue were days of assimilation and absorption, days spent sorting out the sights, sounds, and smells of the unfamiliar. I had never before walked in the Andes Mountains, never stood on the dry and crunchy litter of a pehuén grove or on a bank of a pristine mountain river that ushered along no life. Most of my days I hiked alone collecting the water samples I needed and wandering where my curiosity took me. Actually, *alone* is inaccurate; I had Junior to accompany me.

I still do not know where Junior came from. He found me everyday as I left town to hike up into the mountains or around Lago Caviahue. I developed a strong love-hate relationship with Junior. He knocked me into the acidic water of the Rio Agrio, he dug furiously into the river bank above me, filling my backpack with dirt, he chased after the horses belonging to the Mapuche, he swooped in under my camera lens and made off with an exquisitely shaped pehuén cone I was trying to photograph. He was a pain in the neck. But stalwart hiking companions are hard to find and, in reality, I had little choice in the matter; he did not comply when I asked him to stay behind. Irish setters are beautiful animals but it is not a breed that always responds well to authority, particularly weak attempts at authority. I never managed to sneak out of town without Junior. As to Junior's perception of our time together it appeared evident from the bounds and leaps he greeted me with every morning—he and I were having a hell of a time.

Starting out one dawn, Junior and I spent the day tracing the Rio Agrio up into the mountains. A hoof path worn by Mapuche horses follows along the river a short way. The trail starts to ascend and the river cuts deeper into the hills, away from the path. Junior, I could tell, was familiar with this route and did not care to linger over any view, no matter how startling. Every time I stopped, he barked. I let him stand there and bark, then leap and bark, and finally, whine, snap, and bark, as I studied an amazing intrusion of columnar basalt that formed a plateau on the other side of the river. It was a magnificent sight. This formation with its sweep of columns seemed too fluid, too lyric, to be of rock. Such a dramatic display of inner-earth force—this once liquid earth now stopped in time as stone. Later, I walked across the top of this intrusion and found that the pattern made by the tops of these basalt columns was quite similiar to the textured bark of the pehuéns; nature mimicking itself in polygons of bark and stone. A more reflective landscape for these trees could not exist.

A short distance up river, Junior waited for me above La Cascada del Basalto. Here the path forks down to the bottom of the first of four cascades. When he saw my intention he dashed down the hill in front of me leaving small slides of loose rock in his wake. This waterfall flows over an arch of basalt, trying with time and constant flow to smooth out its raw texture. The pool beneath the fall is such a vivid turquoise that, in contrast to the surrounding red rock and flesh-colored sand, the scene vibrates into the surreal. Not one of these natural elements looked natural to me. The only thing that grounded my visual sense was the presence of a manic, red Setter chasing rocks down the hill. Although the next three cascades were not as unique looking, they were much higher falls and they thundered and splashed to command as much attention. At each cascade: La Cascada Cabellera de la Virgin, Cascada de la Culebre and La Cascada del Gigante we scrambled down to the bottom of the falls, to look, to gather and to chase rocks.

Following the river to take water samples, temperature readings and pH levels, I carried meters and flasks, pencils and graphs, camera and journal, all to record information as to what was there. But standing looking clearly through this water at scoured rock and black volcanic sand, seeing nothing minutely green or growing anywhere near by, I could only perceive what was absent. My concept of water, I realized, had much to do with all that goes on, in and around it. Water teems; it is a giver of life, a flow of living things but here I had to readjust my thinking. Cascading, bubbling, flowing wildly here and there, it was still a beautiful thing but it lacked all that we take for granted water will always be—a sustainer of life.

After the day's water samples were gathered and I sat myself down amid a cluster of pehuéns, it was still absence that I noticed looking out over the Rio Agrio Valley. From where I rested I could see no buildings, no people, there was no wind. I tried to think if ever before I had met with such silence. No highway sounds whizzed and spat in the distance, no planes hummed overhead, no birds sang, no bugs buzzed, not even a dog barked—Junior, impatient and exceedingly bored with my hot and stationary self, had gone back to town. The air was full only of the dry and quiet heat.

Wrapped in such silence with inner and outer distractions

at a minimum I started to examine my immediate surroundings more closely. I spied a petite, gray and orange butterfly moving silently in fits and starts, zigzagging by on some indiscernible course. Then a small lizard darted out from under a rock, dashed up over another, then stopped half way down the back side suddenly aware of my presence. A second tiny lizard rustled by. I caught another movement out of the corner of my eye. I turned as a speck of feathers alighted on the branch of a young pehuén behind me. It looked very familiar, this brown little bird with tail tipped up. It looked a great deal like a house wren. I quickly got out my field guide to Argentine birds and tried to identify what South American species it might be. I looked through the whole book at all the small brown birds. I found it, Ratona común *Troglodytes aedon* — a house wren. I felt rather silly, even sitting there alone and un-witnessed in my mistake. I know North and South America share many bird species and many migrate this far south, but in this strange landscape I did not expect to see such a familiar sight. I did not see it for what I knew it to be. I did not trust my visual sense. Expectation can color perception or worse, close us off to new observations. Nature continually offers a chance to learn this lesson. Sometimes you must trust what you see, not what you know — or think you know.

So far the only bird I had come across in this silent valley was this tiny wren. One wren, a couple of lizards and a butterfly or two were not much to write home about, so after I had filled myself with the quiet of the afternoon I headed upland, toward the volcano. I climbed up what looked to be the highest point in view and found a cairn stacked high. Jumping boulder to boulder up to this pile of stones that stood about my height, I took in the panorama from valley to volcano. The volcano seemed closer, more imposing from this spot. I could see clearly for the first time the fissure below the crater, the gash where yellow, scalding fluid leaves inner-earth and becomes the headwater of the Rio Agrio. My days were ticking by under this volcano and I knew I would soon have to figure out my trek to the top, to the crater lake. In Mexico, many years previous, I had climbed the majestic and then quiet Popocatépetl and in Costa Rica I stood near the base of the booming, ash and lava-spitting Árenal but I had never peered down into a caustic, bubbling crater lake. There is a crater lake in Oregon, a deep and blue and very serene lake, a quiet mountain spot where I once camped. But this lake is nothing of that sort, not one thing like it, except it too is in a crater and gets to be called a lake. After a while, I headed down from my vantage point across a flat expanse of empty sand. Halfway across, the ground suddenly failed to hold my step and one leg sunk calf-deep into the earth. This unexpected plunge was followed by another and then another. I was not on solid ground. Trying to keep a sudden adrenaline spike from manifesting as panic, I carefully made my way toward a rock outcropping and out of this alarmingly unstable stretch of sand the whole time thinking — *where the hell was Junior when I needed him?* Though, on hindsight, it was a good bet that my dear, red friend would have just stood there barking at me to throw him a stone – Lassie he was not.

That evening, sitting with Ernesto and Analia, I went over my day, over my route, asking question after question about all that I saw. My first question, however, concerned the sinking sand. Fear has a way of leaving a most indelible impression. He explained that the Tuco Tuco de maule (*Ctenomys maulinus*) or

"Tuco Tuco," Ernesto nodded and laughed, "like your gopher." Tunduco, as they are called in Mapudungun, live in burrows underground. Then Ernesto, not aware of the extent of my heart thumping panic in the land of the Tuco Tuco, moved right along to what he assumed would really matter to me about this burrowing creature. He gestured to my notebook, "They eat Araucaria seeds." And not just the Tuco Tuco, he went on to add, also the Cachaña or Austral Parakeet (*Microsittace ferrugineus*) which flies, chatters, and squawks through Caviahue in the fall looking for this favored food.

Ernesto had told me the Mapuche also wait for fall and the ripening of the pine nuts before they leave Caviahue and their veranadas, or summer homes, to move down to warmer elevations for the winter. In the spring they drive their small herds of livestock into the area and then back out again every fall. The Mapuche here are known for their horses and horsemanship and I often came across them riding these small but beautiful creatures while on my hikes in the mountains, but without exception they avoided me like the plague, a turn of phrase I do not mean to use lightly in reference to any people indigenous to this hemisphere. The Mapuche, Ernesto told me, believe it is always best to avoid strangers.

Earlier that afternoon under a stand of trees I had come across several fist-size rocks with a scrap of rope wrapped around them. I described this to Ernesto. The people who harvest piñones, he told me, do not always wait until the cones fall on their own. Ernesto did not seem to think that these rock-wielding consumers of pehuén seeds were as much a threat to the pehuén's survival as were the livestock that the Mapuche raise. Pehuéns are very slow growers. The sheep kept by the Mapuche trample, and the goats sometimes graze, on the young plants. A year old pehuén is often not more than four inches tall. I had come across several tiny trees on my hikes, ringed with very large stones — the work of Ernesto's protective hand.

After my questions finally stopped, Ernesto asked his. "Did you find Lago Escondido?" He grinned and pointed out their back door. "No-o-o." I answered slowly, slightly confused since there was only a sheer rock face out their back door. "Aqua dulce." Analia added, nodding. She would show me tomorrow it was decided. In this landscape I wondered how I could miss a lake. Too tired to ask any more questions I accepted the maté gourd Ernesto handed me. I must have been very tired, *fresh water* — it didn't quite sink in. Worse yet, I missed the very obvious clue of its name and why I had not seen this lake, Lago Escondido — Hidden Lake.

I had drunk maté before so I didn't make the face they were expecting as I sucked the tea up through the silver bombilla, an ingenious combination of straw and tea filter. The first sip is always strong and maté is slightly bitter, but it is a taste I associate with comfort. Sometimes maté is offered in a hollow, dried gourd, sometimes in a small wooden goblet or maybe a metal, barrel-like cup but in whatever vessel it comes to you, it comes as a gesture of friendship.

focus

    Learning to find things in nature can teach an important lesson on how we use focus to see or perceive things. Unfettered seeing or pure observation is a wonderful concept, but if something goes unseen because the observer does not know what they are looking at, or for, it is useless. An object must trigger the visual response of focus for us to notice it, otherwise it becomes the visual equivalent of background noise. Bird watchers learn very early on about selective focus. Not only by experiencing those few aggravating seconds lost behind binoculars trying to find and sharpen the image of some zooming missile, feathered and flying by but also by grasping the phenomenon of our own visual ability to separate one thing from another. When new to the field, a birder experiences over and over the frustration of not seeing the bird everyone else has their binoculars trained on. This happens, not necessarily for lack of a dog-eared familiarity with the pages of their field guide. It happens because a two dimensional image on a white page does not give a brain much useful information out in the green, brown, and gray tangle of a bird's life.

    Sometimes focus depends on what we have learned or not learned to see. The event of finding my first bald eagle perched somewhere in a mile of dark green hemlocks taught me much about seeing and focus. One cold winter day, brought to the edge of the Connecticut River, I looked for eagles on the river's far shore. I tried to pick them out along a stretch of the steep, wooded bank but it seemed an impossible feat.

The person standing next to me was scanning the same view counting out one...two...three...four...in an infuriatingly rapid manner. *Where? Where!* I stopped myself from actually stamping a foot. *Look for golf balls,* she answered and kept scanning. Thankfully, she finally stopped and clarified this suggestion. The adult's white head looks like a round, white dot among the dark trees. I resumed scanning the trees, focusing on any small white spot I could find and not, as I was previously doing, looking for something I had never seen before — an eagle perched in a distant hemlock. Damned if I didn't find an eagle almost immediately.

Once I found one golf ball head, I understood the size and shape of the bird I was looking for in relation to its surroundings. Though the white-headed adults were much easier to find, my eyes were now able to focus on any appropriate size blotch on that far shore. Even a few dark-headed, immature bald eagles appeared before me on that shadowy, steep hill. I was looking at exactly the same trees as before; the birds had not moved only now I could see them.

On my way to meet Analia the next afternoon I was still wondering about Lago Escondido. I certainly knew what a lake looked like; how was it I hadn't seen this one? I thought about the fresh water aspect of our destination and what that might mean for the possibility of seeing more wildlife. Since my arrival in these mountains I had scanned the vast sky for any soaring type thing and with great hope of seeing, even at a distance, an Andean condor. But not much large or small had soared, glided or flapped by overhead. In this valley where Lago Caviahue glistens and sparkles blue like any proper picturesque mountain lake should, the dearth of flying fauna or green and soggy flora was starting to unsettle to me.

As Analia and I stood at the bottom of the escarpment out behind their house, I over-played the look of confusion Analia was probably accustomed to seeing on my face. She pointed straight up. I squinted into the sun — I had not thought to look overhead for a lake. I was living in an out-building behind the Hotel Caviahue and this 1,500 foot high escarpment was the view out my window. It was not that I hadn't noticed it was there. Every night a sliver more of the moon would rise from behind this plateau and perch for a moment on top. Every dawn the sun lit the bare rock a pale rose. It blocked my view of the volcano. I had wondered at its geological formation but not once, ever, did I look up and imagine a lake hidden somewhere up there.

I followed Analia wondering how we were going to make our ascent but "ascent" turned out to be too dramatic of a word for our climb. Although it looked to be a sheer face the basalt, eroding and splitting off in foot size rectangular chunks, made for a helter-skelter kind of staircase. Sometimes we had to double back when we ran into a stretch of wall with no footholds but for the most part we climbed unimpeded. Almost at the top, I stopped to look out over the valley. The sky, the mountains across the valley, the lake, even the metal roofs on the now tiny houses below in Caviahue reflected blue. Through all this refracted blueness, the volcanic soil looked purple. Not even one wisp of a white cloud blew by to give depth to this high altitude, monochromatic view.

Standing on flat ground, Analia waited for me as I hauled

myself up the last few feet. She was waiting for me not fifteen feet from the edge of a small lake. She stood there catching her breath and waiting to catch my reaction. Behind me, far below, was a deep blue valley and before me, a snow-shrouded volcano reflecting in a small but shimmering lake. Young pehuéns ringed its shore; there was grass and there were birds. Like the end of any good magic act I laughed in astonishment as Analia reveled in delivering this trick of a lake.

We walked the rim of the lake. The direction of the afternoon sun told me that a picturesque shot, sunset on an indigo lake reflecting a warmly tinted volcano, would not happen that evening. It was definitely a morning shot. I would have to come back up for sunrise. We meandered the rest of the afternoon across this small plateau and down its backside. I did not stop to take many photos or identify all the birds; I was enjoying Analia's company and our conversation and did not want to interrupt the flow. I gave myself the rest of the afternoon off just to saunter. I had already decided on a pre-dawn climb the next day for a few sunrise shots, bringing along a bird book and a packed lunch to further enjoy Lago Escondido and all its hidden wildlife.

That night, back at their house, after expressing my delight with Lago Escondido, Ernesto and I returned to our favorite subject and he presented me with a crash course in pehuén reproduction. Hiking around Lago Escondido I had been keeping an eye peeled for sexually mature pehuéns. Tree species that are dioecious are not common and I wanted to make sure I had photographs of both male and female trees. Ernesto explained that pehuéns do not reach sexual maturity for sometimes up to thirty years. It was fall, so the problem was not a scarcity of cones and catkins the problem was I had not found the right vantage point from which to shoot. My camera lens was not quite long enough for these very tall trees.

The rust-colored catkins of the male contrasting against the green of the tree are the easiest to spot. They are 4-5 inches long and 1.5 to 2 inches in diameter but these branches ending in this dab of color were a distance away. There are 10-20 pollen sacs in the center of this flower and wind must blow this pollen to the cones on the female tree. Considering this on such a sunny and sedate day I pondered the odds of pollination. I had not yet experienced the wild, west wind that blows up and over these mountains. I had not yet been pushed down by that wind or experienced the terrifying lightning storms that sometimes leap over these mountains with it. The evidence, of course, was there with or without the wind. There was no lack of seed producing cones — obviously, pollination was taking place. These impressively large cones, 6-8 inches in diameter, look like giant, hairy artichokes. It is the long, thin scales covering the seeds that give the cone this bristly appearance.

The cones do not hang down from the limbs of the tree but grow up at the end of a branch. The cone- or catkin-bearing branches point up at the ends, creating a candelabrum on which the flower sits. Split one cone open and there are 120-180 nuts to eat and no hard shells to crack one's teeth on. Each of these almond-colored seeds is encased in a thin covering, or pellicle, that has a slight orange cast. The mild and smooth taste is similar to pine nuts eaten elsewhere throughout the world but they could never be confused with any of their small northern relatives; these nuts are often two inches long, which is huge in the tiny world of pine nuts. The Mapuche grind these nuts to make bread, they cook them in stews and roast them to eat. Raw, the seeds are sometimes fed to livestock; fermented, they provide the Mapuche with sustenance of a different and drinkable kind. The Mapuche also collect bags of them to sell to the tourists. In this early part of March, the beginning of the austral fall, many of the cones still remained on the trees. I never found more than a few of these wingless seeds but the ones I found were always in amongst the rocks, very near the female pehuéns. Wind or no, theses seeds, much like the proverbal apple, often do not fall far from the tree.

On one of my first hikes just outside of town I had noticed a small pehuén that had been cut down. Ernesto had told me the trees in the park were protected so I had taken particular notice. It was an old cut and from this severed trunk new growth had sprouted and was well on the way to becoming a new tree. This, I had read, was unusual for conifers but as Ernesto confirmed later not for *Araucaria araucana*. On my afternoon walk with Analia up on the plateau I had also noticed several examples of root suckering, a mode of pehuén regeneration that Ernesto indicated was common in Caviahue. One such instance looked like the Loch Ness Monster appearing out of a sea of sand. The thick, thigh size root surfaced about twenty feet from the parent tree and from it sprang the only green in a large expanse of rock and sand—a tiny new pehuén. In Caviahue, seeds might get some help in dispersal by hungry tuco tuco, cachaña, or rock toting *Homo sapiens* but odds are not great that these species are any more help than hindrance. It is a sound plan that the pehuén has these two alternatives to sexual reproduction.

form

A concise description, one paragraph of form and function, cannot describe a pehuén. *Araucaria araucana* grows very large, very slowly; it lives for many, many years and to describe the tree at only one point in time in this long life, a description could well deceive. Its characteristics and visual description are complex. The pehuén illustrates a visual allegory much like that of the tactile tale of the blind men and the elephant. Ask six people to go out and describe the first pehuén they come across and depending on what stage of growth this tree is in, six different trees might be described. One person might stumble across, almost step on, in fact, a tiny tree of soft leathery leaves. At this stage, pehuén leaves are long and thin and they spiral along the trunk like a twist of DNA. The age of the tree can be determined by counting each complete spiral. This is easy to do since at the beginning and end of each year's growth the leaves are shorter and the angle they grow out from changes. Another person, finding a slightly older tree, would describe a tree that has bark on the lower portion of the trunk and is partially covered with scale-like or imbricate leaves above. These leaves are sharply pointed and there is nothing soft about them. It has a bizarre shape for a tree, almost a t-shape with a few scraggly branches growing horizontally at the top of the trunk. A third person might report back that they find the tree a thick tangle of limbs. At this stage, the tree's wild teenage years, the production of successive levels of branching is more rapid than trunk growth. This is really the only time the tree is densely foliated or takes the usual triangular shape of most conifers. It is possible that one of a myriad of common names for the *Araucaria araucana* — the monkey-puzzle tree — was derived from the appearance of a pehuén at this stage of growth.

Monkey-puzzle is not an English translation of a Spanish phrase, it is not the vernacular of these mountains; there are not, and have never been, monkeys in the Andes. During my first days in Caviahue I tried to figure out exactly what might puzzle a monkey about these tall, towering trees with widely spaced, often pendulous branches. The branches, clothed in their sharp, scaled leaves do not look too comfortable to nestle into, but from a tree climber's point of view I did not see what would be the puzzle about them. A monkey might have to climb a good 100 feet before it reached the first branch, but the bark is so thick and deeply textured that this monkey (presumably with no fear of heights) would more likely just scale the tree than stand beneath it scratching its little monkey head.

In this contrived parable of the pehuén the third person that reports in with a description of the pehuén gets a name — Charles Austin. Charles does not come across his pehuén out wandering the Andes. He is a lawyer in England, a guest at the home of one Sir William Molesworth, and Charles finds his tree at a garden party in Cornwall. This party, in reality, took place in 1834 on the occasion of a ceremonial planting on Molesworth's estate of a "Sir Joseph Banks pine" (*Araucaria imbricata*, the scientific name of the pehuén used in England at the time). On seeing the honored tree Austin is said to remark: *How that tree would puzzle a monkey!* (Hughes, 1982) Perhaps this young tree and its mess of branches would be a tough go for a monkey, had one been put to the test that day, but my guess is that Charles' musing probably ended right there as he toddled off to refill his glass of gin. The name, however, stuck. Sir Joseph, the director of the Royal Botanical Gardens at Kew, if present, might have been amused by Mr. Austin's remark. Of course, at the time Banks could not possibly foresee to what degree he would come to be displaced in *Araucaria* history by Austin's mythic monkey. In Europe and the United States this tree is still often referred to as the monkey-puzzle tree and no longer as the less exotic and rather stiffly formal *Sir Joseph Banks pine* as it once was.

While Mr. Austin in Cornwall is busy refilling his glass with gin there is a fourth person back in the Andes trying to describe a pehuén, gazing, maybe gaping up at a tree that has reached sexual maturity. So far, the first three people in this story have described trees in the earlier stages of development. After becoming sexually mature the trunk continues to grow in height and circumference but if these last three people come back with different stories it is because the crown takes on several different shapes as it ages. In young adulthood, although the pehuén has a trunk that is still relatively thin, its distinguishing feature is a graceful, rather full ovoid shaped crown. Then somewhere around the hundred-year mark, still a very young age for a pehuén, the bottom branches begin to drop off. The fifth designated observer, maybe an artist wandering around with charcoal under her fingernails or ink in his hair, might report back that the crown of the tree looks like a haystack, not a square-bailed stack of hay headed for a New England barn or a giant, round roll of hay wrapped in plastic and left out to winter in a mid-western field but the classic hand-forked picturesque-type mound that Monet found in the fields of France. The haystack-ish silhouette of the pehuén will prevail into adulthood where from then on into old age, which can be an impressive 1,300 or so years, the tree will lose more of its lower branches as it still continues to grow in height and circumference. When the last person reports in, he or she might think they are clever to insist the pehuén looks just like a gigantic umbrella. The droop of those few branches remaining at the top of the tree now give the crown this distinctive and most familiar look. In Argentina and Chile, people often refer to the pehuéns as los paraguas — the umbrella trees. Like the poor elephant being groped by the blind men — six different creatures described from just one poor, manhandled elephant — here six different visual descriptions portray one not so simple tree.

Trying to beat the warm, early light up to Lago Escondido, I stepped outside as night was just starting to move elsewhere in the world. Forms started to take shape in the soft gray, not-quite-dawn when seeing is possible, but focus is unnecessary. It is a time to linger between worlds, to be awake but still shrouded in the haze of sleep, drifting slowly back from the unconscious. The world, hardly seen through this thick gray, is a quiet, cohesive place, a whole, not yet refracted into parts by the sun's rays. I climbed up through this deep gray and was waiting at the top, out of breath but waiting, camera in hand, just as the light came up over the rim after me, scattering the gray and revealing each object around me in a soft carnelian light.

I framed volcano and lake, took a few shots then stood and watched this warm, rosy scene rapidly pale into another day. There were even more waterbirds rippling the glassy blue surface than the day before; I watched Ash-headed Geese (*Chloephaga poliocephala*) honk and preen, Andean Lapwings (*Vanellus reslendens*) and Southern Lapwings (*Vanellus chilensis*) stride along the shore, Speckled Teal (*Anas flavirostris*) and Andean Gulls (*Larus serranus*) paddle across the small lake. I watched tiny fish dart back and forth in the shallows. The cool, fall mountain morning chilled any designs I might have had on a swim, but it did not matter. I sat, then walked, then sat again by the sweet water, watching and listening to life in and around Lago Escondido. I spent the next few hours wandering across the top of this plateau. Here there were more young trees than I had come across anywhere else. These trees appeared a tangle of woolly branches with each limb covered completely from base to tip in a spiral of thick triangular needles. To refer to them as needles is misleading; these structures look more like a whorl of agave cactus leaves spiraling around each limb. It is only at the point does the word needle seem appropriate. At a distance the limbs look soft and thick, sheathed completely in these layered green leaves, but up close and in contact with the real thing, the tips are sharp — very, very sharp.

It is difficult to walk silently beneath a pehuén. It is not the forest floor of a white pine grove where footfalls are swallowed by a deep cushion of pine needles. The debris below a pehuén crunches and cracks loudly underfoot. A faint dry spice of a scent lingers around these trees; it is hard to put a nose to, hard to name, but walking underneath them stirs up a drift of earthy perfume. The resin of the pehuén burns with a stronger version of this dry and distant smell, a phantom smell, starting somewhere in spice then trailing off into faint pine. It smells wonderful and burns well. When the Spaniards showed up, bringing with them their penchant for masking smells and for smokey rituals, they used it as incense, comparing it to the familiar scent of another ever-popular tree resin—frankincense. (Molina, 1809) According to the *Journal of Ethnopharmacology* the Mapuche have used the sap for medicinal purposes, applying the dark amber-looking ooze of the pehuén to help wounds and bruises heal and to relieve headaches. (Houghton, 1985)

This dry leaf litter beneath the tree is also thick with branches. Standing below a mature pehuén, looking up, the tree looks like it has been limbed. It has not; the branches drop, fall completely off, after they stop producing cones or catkins. Once the tree is sexually mature the branch growth for one year forms on a single plane on the trunk, producing anywhere between three to eight branches in a whorl. The wood is hard, especially for its light weight but the *clavos* or gnarls at each node of a branch are even harder. If the trunk is cut in a cross section through a whorl, the deep golden color of these clavos protruding into the white-blonde grain of the trunk creates a striking pattern, a sunburst of gold. Sometimes this pattern shows up when a tree is cut but these clavos are tough on any saw. They are avoided for another reason, Ernesto had warned me. The gnarls when burned in a wood stove burn hot, sometimes too hot. Stoke a stove full of these and the burning clavos, he says, can melt the sturdiest of caste iron stoves.

After a morning of bird watching and aimless wandering around Lago Escondido I turned to stalking less mobile prey— examples of the reproduction and life cycles of the pehuén.

I had hoped to find the unusual instance of a pehuén that is monoecious. There are references to their existence but in all my hikes in the area I never found an example of this tree with both male and female flowers. (Veblen, et al. 1995) Identity is not a singular thing for the pehuén and to describe *Araucaria araucana* as a dioecious species does not really explain this tree. Dioecious, monoecious, regeneration, root suckering—all are aspects of continuing on as a species—focus on one and the others blur in importance. One might be more prevalent then the others, more obvious, but many alternate modes of reproduction exist and that is an important part of the identity of this tree.

As I came down from Lago Escondido and headed toward the hotel, Junior caught sight of me from the road. He leaped and jumped at me, trailing off his deep bark with a high-pitched whistle. He was not wagging his tail. I tried to apologize for ditching him as I dodged his leaps. He nipped at my sleeve; he was one mad, red dog. Finally, he calmed down and ran along in front of me toward Ernesto and Analia's house. Ernesto had known my plans for the day and wanted to know how I made out photographing the pehuéns. He nodded as I ran down the list. I had told him I would send him copies of my slides to use in his work, so he listened closely to my descriptions. He nodded again when I finished, but his eyes told me he was waiting for something I had not said. The male and female, he wanted to know, did I see the difference? I started in on cones and catkins and he shook his head. "No," he smiled, "without those."

Blankly, I stared back at Ernesto. He took my pen and started drawing. Some branches, he explained, grow differently: male–straight, female–zigzag. I could tell he was enjoying imparting these particular details. He drew two stick figures and labeled the trunks 'I', the main branches 'II' and a branch off each main branch as 'III'. He explained that it is on these secondary branches 'III', that the tree bares its cones and catkins and it is here, even without these structures present, a female can be discerned from a male. A year's growth on one of these secondary branches ends with the production of a cone or a catkin, depending on the sex of the tree. Come fall, this structure does just that, it drops off and the following year this same secondary branch will grow another segment and produce another cone or catkin. This process will be repeated for between four to six years then when a secondary branch ceases to produce a sexual structure the whole branch dies and falls off. Until then, however, the succession of each year's new growth and the branch structure it creates is a visual clue to whether one is looking at a female tree or a male tree. Both cone and catkin in the first year are produced from a terminal bud, but after that male and female bud formation is different. On the male (fig.1) it again forms at the terminal bud and the branch continues to grow in a straight line. But on the female (fig.2) the second year cone forms on a lateral bud, to the side of where the last year's cone was formed. This causes a visual change in the branch structure, one year's growth zigs to one side and the next year zags to the other, no straight line. And no, I hadn't noticed.

A sexually mature pehuén is a tall tree; the branches producing the flowers are a good distance away — but still. My ability to adjust focus, to learn to see what I was not looking for, had failed. I could not lay blame on the tricks of the sun;

these branches were not hidden in shadow, shrouded in gray, they were not obscured from view. My focus was diverted by assumption. In looking at both the male and female pehuén I saw only the difference in the structure of their flowers. I knew nothing of a dioecious species whose branch structure differed in the sexes. I missed what was right before my eyes, didn't see the branches for the tree, flunked the observation test. But it was a bigger problem than just focus; in looking at the form as a whole I had over-looked its individual elements.

figure 1. male

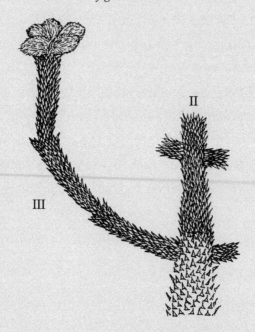

figure 2. female

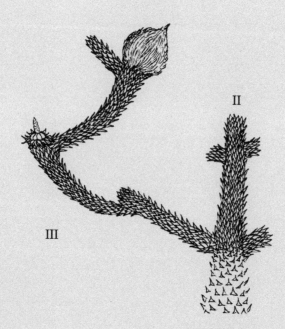

shape to word

In nature, shape often tells us what we see: the oval crown of a sugar maple, the flounce of a weeping willow, the V at the tail end of a barn swallow. Shape and form help us identify things. It is how we can discern an eagle from a vulture, a thousand feet straight up, unique black shapes against blue sky. Looking at the shape of a leaf, the tiny outline of a bud, or the silhouette of the entire tree, are all ways to help tell one species of tree from another. Studying shape, form, and line can be a way of seeing a whole *and* its parts. Learning well the characteristics of a species allows for rapid identification, however, sometimes visual memory substitutes shape for object. When this happens the eye identifies a familiar shape then moves quickly on, sure of what it has seen. In nature, categorizing by shape is an effective memorization tool but when it becomes habit, a function of knowing, not seeing, we are more likely to miss something, less likely to learn anything new.

It is the same with words. Substitute word for object and the full nature of an object sometimes gets lost or is never fully explored. Jorge Luis Borges took this concept a step further, he maintained that any linguistic construct was a temporary ordering of things, a fiction of sorts. In truth, we have learned throughout history that science, like language and its constructs, can also be a temporary ordering of reality. Facts are facts but they do come and go. But a fact, whether it is on its way in or out, can be important to note and these linguistic constructs, fiction or no, might just have a story to tell.

Araucaria araucana has more than its share of common names, but then, it is an unusual tree: pehuén or pewen, pino cordillerano, pino de Chile, pino de la tierra, los paragues or umbrella tree, Sir Joseph Bank's pine, monkey-puzzle tree—all these names tell something of the tree or its history. *Araucaria araucana* is its scientific name, a name unique to this species but in keeping with the unusual nature of this tree it has not been a species unique to one name. It has had at least eleven. Eleven scientific names, one tree; this reveals something about the pehuén—this tree was not ignored by botanists.

Disputes, heated discussions, even some not-so-polite name-calling took place in Europe during the 18th century but not just over the pehuén. The age of what is called *enlightenment* in Europe was in full swing and there were quite a few very serious arguments taking place at this time in philosophy, religion, and certainly, the sciences. In the natural sciences, as Linnaeus' system of plant classification was in the process of being hashed out, botanical collections were moving from cabinets of the curious to the focus of serious study. European expeditions were being dispatched around the world for the purpose of plant collection, or resource piracy, depending on the viewpoint of where one stood—on the deck of a foreign vessel or on a beach in this new world. For medical and economic reasons, what was called the *New World* flora held a point of much interest for the Europeans. While all this was going on, the pehuén, this tree from Patagonia, somehow got caught in a crossfire of budding botanists, muscle-flexing monarchs, and the politics of the Catholic Church.

Abies araucana, Abies columbaria, Pinus araucana, Pinus chilensis, Dombeya chilensis, Dombeya araucana, Columbea quadrifaria, Columbea imbricata, Araucaria chilensis, Araucaria imbricata, Araucaria araucana – each one of these names has been published as the scientific name for the pehuén. If scientific names are chosen to be unique representations of a species there seems to have been a good amount of confusion surrounding the pehuén. Some thought it should be placed in with the firs (*Abies*), others the pines (*Pinus*) and those botanists who realized it was a new genus altogether had different ideas as to what to name it or who to name it after. It took just shy of a hundred years to sort things out, to delineate who published what, when and to decide why the pehuén should be named *Araucaria araucana.*

A German botanist, Karl Koch, in 1873 wrote *Dendrologie,* and it is, as the title implies, a study of trees or more accurately, *a study of* the study of trees. In listing the pehuén, Koch waded through the long list of scientific names in an effort to place order on the chaos of this classification. His efforts led to verification of who published first on the pehuén and which nomenclature was most valid. Now, a century later, Koch's name can still be found attached to this scientific name to summon up, it seems, the validity of the provenance of this name. Because of this, a reference to the pehuén can take up quite a bit of ink: The pehuén, *Araucaria araucana* (Mol.) K. Koch — in other words, the common name, *the scientific name* (the guy who named the species) and the guy who says that this is the guy who named the species. It all seems pretty clear.

Juan Ignatius Molina, a Jesuit priest born in Talca, Chile, is the man in parenthesis, the person Koch puts forth as the first to publish on the pehuén. The need still to clarify this title may stress the amount of confusion generated around the pehuén and the discovery of the new genus it represented. It also might be indicative of how deep territorial rights to this name ran. Molina, born on Chilean soil, was the only South American in the gang of botanists to publish on this tree. The more important fact, though, is that he was the first. This genus of tree found only in the Southern Hemisphere, a species native to Chile and Argentina, had most of its turbulent taxonomic history hashed out in Spain, France, Italy, and England but did end with a Chilean botanist getting the credit. Because this botanist was banished from his homeland, the saga does not end in Chile — but it does begin there, in an appropriately named place for the beginning of anything — Concepción.

If a movie were to be made of this story, a cinematic version of this taxonomic tale, the opening scene would have to be a shot of a dark and rainy indigo sky lit up by a bolt of yellow lightning. The crack and thunder of lightning would mix with the sound and then the sight of a splintering, burning mast. The camera would pull back on the Spanish warship, *San Pedro de Alcántara*, being tossed about in the harbor of Concepción, Chile and a slow motion pan would follow the flaming mast as it falls and splashes into a furious sea. Events would then unfold, portents and all, from this opening scene tangling the pehuén up with economic interests in Europe, a slew of botanists and eleven scientific names.

That splintered mast on the warship *San Pedro* was

replaced by a pehuén. A year later, in 1782, this mast, and stories about the tree it came from, drew the interest of three visiting botanists: Joesph Dombey, Hipólito Ruiz López and José Pavón Jiménez who were trekking about Peru and Chile as part of the Spanish based *Expedición Botánica of 1777-1788*. They had arrived in Concepción to set up temporary headquarters for one of their forays across Chile and planned to stay awhile. The desire to see the tree that this tall mast was cut from lured one of the botanists, Pavón, out into Mapuche territory. It was, so the chronicle reports, a brave thing to do. Accompanied by a naval officer into what they called hostile territory, Pavón brought back a few branches with cones and catkins attached. The three were convinced they had found a new species of what they referred to as the Linnaean genus *Pinus* and as Ruiz states in his chronicles: *We agreed unanimously that it was a new species of this genus and probably the most valuable of all those discovered up to that time for its exquisite white wood of excellent grain.* (Ruiz, 1940 ed.) The value of the pehuén, they could see, was in its tall, straight trunk and dense, light wood. At the time, the European navies were all lusting after any seafaring advantage they could find. The competition to survey/control/exploit this new land was fierce and sea power, among the constantly quibbling Europeans, was a very serious matter.

The idea for the expedition that brought these three to Chile had begun in France. Louis XVI's chief minister, Monsieur Turgot, asked Charles III of Spain (Louis XVI's uncle) if they could send a French botanist to what they called Spanish America. At that time, France had yet to create a collection from the prolific South American continent and was eager to do so.

Botanical research had been going on for some time in Paris, in the middle of which was usually a member of the Jussieu family. Three brothers, Antoine, Joseph, Bernard and their nephew, Antoine-Laurent Jussieu, were, in effect, a dynasty of botanists that held sway over plant taxonomy in Paris for close to a century and a half. When the idea for this expedition surfaced, Antoine-Laurent suggested that Joseph Dombey, the médecin-botaniste at the Jardin du Roi, make the voyage.

Charles III of Spain was a relatively enlightened guy who supported scientific research and the botanical arts. He conceded to letting Dombey study in *his* colonies on the condition that Dombey be accompanied by two Spanish professors and that he also agree to leave a duplicate of all that he collected with Spain. It took the Spanish crown awhile to get the expedition's wheels in motion and they finally chose Hipólito Ruiz López and José Pavón Jiménez to represent Spain's interest. Both had a background in pharmacy and compared to Dombey had little practical experience in botany. Dombey, Ruiz, and Pavón along with two artists, Joseph Brunete and Isidro Gálvez made up the *Expedición Botánica of 1777-1788.*

It is difficult to pinpoint when the trouble between Dombey and his companions began but it seems it was inherent from the start. In a letter to a colleague Dombey proclaimed that Spain was to provide him with two *pupils* as his traveling companions. (Steele, 1964) Although Dombey was certainly the more professional and experienced botanist he could not, from the Spanish view of things, head up an expedition of Spaniards in Spanish territory. Hipólito Ruiz was appointed first botanist. The three apparently managed to get along well enough for the many years they trekked throughout Peru and Chile, but after six years Dombey decided to return home ahead of his companions. Complaining of health problems from the effects of scurvy, his eyesight and hearing diminishing, Dombey petitioned to leave the expedition and return home. He had signed a contract not to publish without Pavón and Ruiz but his early departure leaving Ruiz and Pavón still out in the field made the two Spaniards nervous.

On his return to Europe in 1785 after a very rough voyage, Dombey was forced to spend a great deal of time in Cádiz trying to get his specimens out of Spain and home to Paris. He argued long and hard that the collection of Ruiz and Pavón would indeed be the duplicate of his own which the Spanish crown required and that most of it was already on its way back to Spain aboard the *San Pedro de Alcántara.* To release any of his specimens, among which were some seeds, bark and cones of the pehuén, the Spanish officials insisted Dombey sign an agreement not to publish without Ruiz and Pavón. Words were exchanged between France and Spain while Dombey waited it out in Cádiz slowly amassing more and more debt. In a letter to Gómez Ortega, the official who held the fate of the Frenchman in his hands, Dombey declares that he will never get to Paris and is resolved to die in the poorhouse. (Steele, 1964) Despite this dramatic pledge Dombey did submit to the apportioning of his cargo by Spanish officials and returned to Paris with only a fraction of his material. Ruiz and Pavón ended up staying on in South America for another five years.

Dombey upheld his promise not to publish but not long after his return other French botanists wrote about his collection, including the pehuén and the new genus of tree they thought it represented. The naming of this genus and the assigning of a species name to this tree started a long and heated debate. The three botanists who had collected the specimens were not in a position to publish—Ruiz and Pavón were still in South America collecting and Dombey was forbidden to do so and that left a publishing opportunity wide open for any botanist who had access to an example of this species. One such person was Chevalier de Lamarck who in 1786 assigned the genus name of *Dombeya* to the pehuén in acknowledgment of the existing but unpublished work of Dombey. In 1789, Antoine-Laurent Jussieu in his *Generaplantarum* counters this designation. Jussieu, like Lamarck, believed it was a separate genus but for some reason he did not go along with the choice of honoring Dombey, even though Jussieu was the one to recommend Dombey for this expedition. He gives the genus the name of *Araucaria* in acknowledgment of the area in Chile the tree was thought to grow.

Loyalty to one's countrymen seems to be no factor in who sided with whom in this debate of designation. When Ruiz and Pavón returned to Spain, Ruiz put together a dissertation on the pehuén trying to gain admission to *The Academy of Medicine.* Ruiz chose not to identify it as a new genus but instead place the species in with the pines. Pavón, instead of backing Ruiz, his traveling companion of eleven years, ditches his fellow Spaniard and sides with Jussieu. Another well-known Spanish botanist, Antonio José Cavanilles also backs up Jussieu, making Ruiz furious. But there was bad blood already between Cavanilles and Ruiz. While Dombey, Ruiz, and Pavón where still abroad dealing with the rigors of expedition life, Cavanille wrote about a class of flowers, the *monadelphia*, from the specimens that the trio had already sent back to Spain. He gave three of these flowers the genera of *Ruizia, Pavonia, Dombeya*. Cavanille appeared to believe he was honoring the three botanists. They were not thrilled. All three had already picked out a favored New World tree to dedicate to themselves and their work — but none of these trees now bear any of their names. After Ruiz and Pavón returned to Europe they realized they were to be immortalized solely by the *minor* flowers that Cavanille had chosen for them.

For Dombey, even after his return to France, life did not go well. First, there was the reality that all his years of work in South America was not publishable. Then he found himself most inconveniently surrounded by the French Revolution. In 1793 he asked for a commission to travel to the rather newly united States. It probably seemed to him a good time to get out of town. Unfortunately, he never quite made it. His ship landed on the French island of Guadeloupe to ride out a storm and Dombey somehow managed to become embroiled in a confrontation between partisan and old regime forces. After recovering from a fever precipitated by an unexpected dunk in the sea, a result of this altercation, Dombey set out once more for Philadelphia. He didn't get very far; British pirates intercepted his ship. When they brought the ship into dock on the island of Monserrat, Dombey, who had previously disguised himself as a Spanish sailor, was carried off the ship and imprisoned. The chronicles say he then took sick and died only a few days later, still in prison.

Things for the two Spanish botanists did not go much better. Early on during their expedition the botanists sent their first shipment of specimens home on the Spanish ship *Buen Consejo*. That ship was captured at sea by the English who sold off the botanists' collection in Lisbon along with the rest of the ship's cargo. Eventually, through much political wheeling and dealing by Spanish officials, the specimens made it back to Spain, however, Pavón, Ruiz and crew on hearing of the initial loss of the ship, had already set out north from Lima to recollect the plants they *thought* were lost forever. Then in 1784, while they were still in Peru, the pehuén-masted *San Pedro de Alcántara* set sail for Spain. On board was a shipment of fifty-five cases of specimens, over a thousand plant drawings, samples of minerals and other miscellaneous artifacts — about five years worth of collected materials. Thirty-one tubs of various live trees were also on board — but not for very long. In a wild storm off the coast of Chile the ship lost its cargo of trees, swept overboard or possibly deep-sixed in an attempt to save the ship. Pavón and Ruiz were still in Peru when word of the storm reached them. Ruiz had already sent some seeds of the pehuén back home but I imagine he regretted deeply that there were seventeen small pehuéns now at the bottom of the sea. And, of course, as bad situations sometimes do — it got worse. Two years later, still trying to make its way home after several other seafaring mishaps, the *San Pedro de Alcántara,* the first ship with the mast of a pehuén, sank off the coast of Portugal — so very, very tragically close to home. The botanists' portion of the cargo, years of work, was never retrieved from the ocean floor.

Later shipments did get through and the boat Ruiz and Pavón sailed back on, *The Dragon,* arrived safely. Cargo, botanists and one of the artists, Gálvez, landed in Cádiz in

September 1788. The other artist, Brunete, never made it home. Ruiz points out in his chronicles that in an unfortunate twist of fate the place Brunete died of pneumonia, Pasco, Peru, was the town the man had hated most in all his travels. (Ruiz, 1940 ed.)

Fortunately, the Spanish botanists were at least able to publish part of their years of labor. First they created the *Prodromus* (1794), which was meant to be the introduction to the many volumes they optimistically thought would soon follow. This publication, a very small percentage of their collection, offered 136 new genera, all classified by Linnaean characteristics. It had been published quickly, since many of their discoveries were already being written about by other botanists and subsequently, many corrections had to be made and entries amended when they began publishing the more in-depth study *Systema vegetabilium florae peruvianae et chilensis* (1798). One of these changes was the re-classification of the pehuén by Pavón from the pine genus into a genus of its own, naming it *Araucaria imbricata*. This huge project, however, proceeded very slowly and more money needed to be found if the rest of the work would ever see the light of day. The Crown, a little short on cash at the time, decided the funds for this project would have to come from money collected from the colonies. Ironically, as it turns out for the Chilean Jesuit Molina, Ruiz and Pavón were told that while waiting for these funds to come pouring in they could use assets obtained from the confiscated properties of the Jesuits.

Pavón was already personally in dire financial straits when in 1808 the French army entered Madrid and Napoleon started ransacking Spain. Pavón began selling off parts of his collection to other European botanists and though Napoleon did not hold onto Spain for very long, when an absolute monarch reappeared in the form of Ferdinand VII, Spain took a few steps backward. Spanish scientists found it tough going under this less than enlightened King. Ruiz, the driving force behind the project, was still pushing to get the rest of the work published when he died in 1816. After his death the project came to a halt. Several attempts were made over the following years to publish more of the work, but it never happened. In the end, only three volumes were produced from this vast, but ill-fated, collection.

England was also very busy at that time out scrambling around the New World collecting things. France and Spain were not the only countries interested in the pehuén. Not long after arriving in England the tree became rather a status symbol, an odd and interesting conversation piece; all the *best* estate gardens had one. Unlike many exotic, often tropical imports, the pehuén grew well in English soil. It seems Archibald Menzies, a Scottish surgeon, had also been to Chile. A botanical collector, he had been sent by Sir Joseph Banks to accompany Captain Vancouver on his *Voyage of survey and exploration aboard HMS Discovery*. He returned to England in 1795 with five small pehuéns. The story is that Menzies, at a dinner party given by the Viceroy of Chile for the officers of the *Discovery*, pocketed a few of the hors d'oeuvres — several very large pine nuts. He successfully sprouted these seeds and tended the trees for the remainder of the four and a half-year expedition, planting all but one in the Kew greenhouse on his return to England. The English, putting in their two cents as to what this species should be called, opted for two different but equally inappropriately popular New World misnomers: *Columbia quadrifaria, Columbia imbricata.* Labeling this towering tree from the Southern Hemishere with the name of Christopher Columbus apparently made sense to someone.

This new genus of tree created quite a ripple in the small pool of European botanists that existed at that time but before any of these expeditions had even *arrived* in Chile a Jesuit priest by the name of Don Juan Ignatius Molina had studied and well-documented the pehuén. Molina had spent much of his life studying the natural history of Chile. His apparent love of Chile and its terrain was evident; he wrote on all aspects of life around him. But Molina was not there to greet his fellow naturalists or to share this knowledge when these other botanists hit shore. Life in Chile ended rather abruptly for Molina and one day he found himself headed out across the sea, unwillingly, probably despondently, sailing toward Europe. The same attitudes that were the driving force behind the botanical exploration of the New World, the importance of analysis, reflection, and reasoning in regard to all things physical and spiritual—the tenets of the *age of enlightenment*—were also some of the same forces that exiled Molina from Chile, separating him from his intellectual pursuits. It was not a good time in history to be a Jesuit.

The Jesuits, the educators of the ruling class, the court confessors, the traditional pipeline to papal Rome, did not fare well during the enlightenment. In the 17th and 18th centuries there were many philosophers: Descartes, Hobbes, Spinoza, Locke, Hume, Voltaire, Diderot, Rousseau that had few kind words to expound on any tradition that held itself above question. A need to place as much distance as possible between this new world of the intellect and a not too distant medieval past was strong. Much of the Catholic Church itself, outside of Rome, was trying to separate itself from recent history and the very bloody, messy business of the inquisition. It was not, as history has shown, a very effective way to coax anyone back into the fold. Meanwhile, even support from within the church was on shaky ground. In the corridors of the Vatican the other orders of the church whispered to the Pope about the Jesuits and their abuse of power. The rulers of Portugal, Spain and France, one by one, ousted the Jesuits from their borders. It wasn't so much a matter of separation of church and state as a separation of Rome and Crown. The absolute monarchs of the seventeenth century all had Jesuits in court — Jesuits as educators — and this made for very close ties to Rome. The eighteenth century was a new age. It was, apparently, time for a change.

In 1767, stripped of his life, his collections and his manuscripts, Molina left for Europe to take up residence in Italy, the only place where he was welcome at that time. While Dombey, Pavón and Ruiz were still in Chile, Molina somehow managed to recoup his information and publish his natural history of Chile, *Saggio Sulla Storia Naturale del Chili*. How this transpired is unclear but the *anonymous* American writer who translated Molina's work in 1808 explained in an intriguingly vague manner that Molina had the "good fortune to regain by accident" his manuscripts on Chile. The only detail given is that the manuscripts came to him via Peru sometime after he arrived in Bologna. This translation, as it turns out, was written in Middletown, Connecticut by a now no longer anonymous Richard Alsop. How Molina's work came to Alsop's attention is not clear but he had both a copy of the work in Italian and Spanish. Alsop was a local poet and one of a group of men known as the Connecticut Wits, or Hartford Wits, a group of writers and poets active between 1770-1810. Alsop, referred to as a "universal scholar", was a writer, linguist, editor, political observer and historian who had a long-standing interest in natural history. Another English edition was published in 1810 in London and, oddly, this too was translated anonymously.

Molina refers to the pehuén in his first edition published in Italy in 1782 as *Pinus araucana* and the same entry appears in the English translation of the 1787 edition:

"The pehuén (*pinus Araucana*) called by the Spaniards *pino de la tierra*, resembles the fir more than the pine, although in some respects it differs from both."

Molina must have subsequently read other material published on the pehuén and humbly assumed he would not hold any influence in the esteemed world of European botanists. In the 1810 edition of *Saggio Sulla Storia Naturale del Chili*, Molina sets aside his own designation and modestly cedes the scientific name for this tree to both Pavon and Lamarck. This time his entry for the tree reads:

"The Pehuén, so-called Pine of Chili, *Araucaria imbricata* Pav.; *Dombeya chilensis* Lam.; the most peculiar, the most beautiful and the tallest of the trees that the Chilean terrain produces."

He continues, explaining his earlier classification of the pehuén, pointing out that he too thought it should be a separate genus:

"It is certain that the tree through its peculiarity deserves well to make a genus apart and willingly I accept and adopt that it be formed under the name *Araucaria*, because it indicates its origin. Finding myself in the Country (Chile), I had done the same, as is seen in my manuscripts, in which this tree is put under the genus *Pehuenia*, thus called by the name given by the Chileans, but having gone back to Europe I joined it to Pines in order to avoid the note of wishing to multiply unnecessarily the number of genera, which the vegetable kingdom of Chili offers."

It took quite awhile for the Europeans to let go of their designations of choice: *Abies araucana*, *Abies columbaria*, *Pinus araucana*, *Pinus chilensis*, *Dombeya chilensis*, *Dombeya araucana*, *Dombeya chilensis*, *Columbea quadrifaria*, *Columbea imbricata*, *Araucaria chilensis*, *Araucaria imbricata* and the confusion of names for this tree lingered on for quite some time. Koch certainly gave credit where credit was due for the species name *araucana*, but had he gone further and adapted Molina's original choice of genus, *Pehuenia* the scientific name would have been a more accurate designation. The word *Araucaria* was a word from the New World, even a word that has its origins in Chile, but it is a word constructed by the Spanish and it has little to do with these trees.

ARAUCARIA IMBRICATA *Pav.*

Chili   Plein air

SE-QUO-YAH.

Philadelphia Published by Key & Biddle

From Childs & Inman's Lith Press.

words

Nouns are observable. To point to a tall, woody, maybe willowy, perennial plant and have someone respond—tree, árbol, aliwen, is a simple thing, an easy communication, the visual transformed directly into language, any language. It is relatively easy to accumulate words in this point and shout manner, there is an object and here is what it is named—no further information is truly necessary. Knowing the etymology of a word at this stage of learning makes little difference. What might be the more important thing to know is from *whom* the word comes; what is the background of the person standing there presenting definitively that word *tree*. Certainly, when information is being exchanged in the same language, questions can be asked, backgrounds accounted for, clarification made, but trouble can arise when asking for words in an unfamiliar language. Important nuances can be missed.

By stopping a person on the street, pointing politely at a particularly tall and majestic tree and asking in Spanish, "What is this?" Instead of the word árbol (tree) the more specific word, roble (oak) could be the answer. If the person happens to be a botanist, the word delivered might be *Quercus*. Arbol, roble, *Quercus*—all labels for the same thing, yet two of these answers are really misinformation. Though somewhere down the road, maybe down a breezy, cool and tree-lined street a person might curiously insist on calling every sycamore, casuarina, ceibo—an oak, little more than mild embarrassment is the result. These kinds of misunderstandings generally sort themselves out rather quickly, but there are some words, other labels, which take a little longer, hundreds of years even, to understand there has been a misunderstanding.

Nouns are labels, labels get co-opted, misappropriated, carelessly used — it is the way of the world. It certainly was the way of the colonizers of the Western Hemisphere. It started when Columbus sighted land and those waves he rode to shore on turned out not to be pounding the coast of India as he first believed them to be. He eventually figured this out but all those who followed after him cared little to correct his mistake of referring to the people of this New World as Indians. Contact with the Europeans stuck those indigenous to the Western Hemisphere with many words, lots of labels, not of their own choosing.

Words written down again and again in history, in novels, in a hundred arrow-strewn Hollywood scripts, tell us that groups of people living in the northern Great Plains of what is now the U.S. are called the Sioux. And yes, it is true, they were called this — by French traders. The traders shortened the word from *nadouéssioux* — certainly easier to pronounce. But, and this is the important thing to know, the French got this name from the Ojibway. The Ojibway, who the French were trading furs with at the time, pointed out the people who called themselves the Dakota, Lakota or Nakota and repeated the Alqonquian word *nadowe-is-iw* to the French. It is said that the word means something akin to snake. Apparently, there were not neighborly relations between these two sets of *New World* neighbors.

Names of people, places, things, all throughout the Western Hemisphere are a mix of indigenous words, words from the culture of the colonizers, words transferred from previous associations those colonizers had with other cultures and hybrid words made from a mix of all these circumstances. Often the source of a word is obvious. In the United States there are mountains, rivers, lakes, regions, which still retain the original indigenous names. These names, however, often get co-opted for other use: municipal titles, names of products, sports teams, the local liquor store and these words are treated as if they carry no meaning for the culture that they have been taken from. But, of course, they do. These words are often appropriated with little understanding as to what they actually mean.

Chargoggagoggmanchauggagoggchaubunagungamaugg is the name, or one version of the name, of a lake in the New England state of Massachusetts. Though the spelling and the exact translation vary, in Nipmuc, an Algonquian language, whether the word describes *fishing boundaries* or *a lake divided by islands*, the word is all about being a lake. It is a place name that is difficult to carelessly use. The present inhabitants don't pronounce the name much, they refer to the lake as Webster Lake but they do bother to squish this long lake title onto a local road sign and across a more accommodating and lengthy railroad trestle. In what is now called the town of Webster, where this lake sits, t-shirts are sold with the name of this lake wrapped around the portrait of a Native American. The present locals do not seem too concerned that the Nipmuc tribe did not dress in the elaborate headdress the figure is depicted wearing; it is not the garb of any eastern tribe. The t-shirts and this long indigenous name are for the tourists but for the obvious length-related reasons this indigenous word retains its meaning as a place, a lake. It is not the name of a coffee shop, a laundromat, nor does it appear on the label of any consumable product. There is no Chargoggagoggmanchauggagoggchaubunagungamaugg beer — yet. The word still remains in proper context.

The redwoods, *(Sequoia sempervirens)* and the giant sequoias *(Sequoiadendron giganteum)* are massive things, very big trees, and as the story goes Stephan Endlicher (1804-1849), a distinguished Austrian botanist and linguist, named these genera after another well known linguist of that time, a Cherokee. The Cherokee *were*, of course, an East Coast tribe, and these trees grow tall very far out west, but that has nothing to do with the inappropriate use of this Cherokee name. Our country is strewn with honorifics coast-to-coast: mountains, dams and highways having absolutely nothing to do with the person whose name they carry but it is the origin of the name, not the origin of the man himself that maybe should have been considered before transferring this moniker to these trees.

When this Cherokee, who came to be called Sequoyah, lived it was not anything close to a good time for the Cherokee Nation; they were being forced off their land and driven out of the East. Sequoyah (1770-1840), a Cherokee Chief, was of the mind that the Cherokee should go along with the seemingly inevitable forced resettlement plan. Not a popular stance with many of his tribe — but this is not what earned him his nickname. His Cherokee name was Ga-la-ge-noh, his English name George Guess and it was only after he became obsessed with inventing a way of writing down the Cherokee language did he acquired the nickname Seqouyah. It seems George met with the suspicion of his neighbors and the distrust of his wife when he started working on a visual representation of their language. The story is that his wife burnt the first set of symbols he had carefully drawn on pieces of bark—to her the symbols were foolishness, or worse, maybe spiritually dangerous. George's response was to move out, build a more remote cabin in the woods, find a new wife and keep working. Certain that this guy was up to no good in his new secluded home, his neighbors started calling him Sequoyah. To the Cherokee *sequoyah* is that small, rather furtive creature that in English is called an opossum.

Sequoyah eventually became the editor of the *Cherokee Phoenix* and he did end up creating a Cherokee Alphabet but after the torturous displacement of the Cherokee nation to Oklahoma from their East Coast land and life, he was assassinated. For his neighbors, Sequoyah might have seemed an appropriate nickname for this odd acting character, however, by transferring this word to a genus of tree and not understanding its origin, this botanist, Endlicher, created a rather strange nomenclature for these massively tall trees. Recently, a Northern California historian has come up with his own version of what Endlicher had in mind when naming these trees and leaves Sequoyah out of the picture altogether. His quest for logic over lore is understandable but logic does not always play a part in the history of words.

Botanists get to name things; this is one of the perks of the job. There exists enough Latin in the world of scientific names that an assumption might be made that the root or exact meaning of a word is an important part of choosing a word to label a species. Often it is, and when the European botanists were arguing over just what this new South American genus of tree should be named, the seemingly sensible idea of using words that have to do with its place of origin won out. *Araucaria araucana* certainly seems more appropriate than *Abies columbaria* or *Columbea quadrifaria* – *Abies* is the genus fir and Mr. Columbus never saw the forests of Chile. The problem, however, with *Araucaria* is in the origin of the word.

The indigenous people that live in the Andes of Chile and Argentina are often called Los Araucanos – but it is not what they call themselves. The different groups that share the same Mapudungun language are Mapuche. In 1569, Alonsa de Ercilla published the epic poem *La Araucana* glorifying the war the Spanish fought against the Mapuche of central Chile. Ercilla portrayed, in his endless length of a poem, lots of battles, brave deeds, and a lovely and tragic Mapuche maiden. These brave deeds were carried out, not only by Ercilla and his fellow Spaniards, but by the Mapuche as well. They are, he tells the reader, a fierce and brave enemy. The Mapuche were apparently fierce opponents and the Spanish, to validate their heavy loses, thought it best to make it known what they were up against. Ercilla, in the beginning of this poem, is obvious about his motives:

> .... *and feats that aggrandize the Spanish name*
> *For the brave actions of the vanquish'd spread*
> *The brightest glory round the victor's head.*
>
> *(Molino, 1808 ed.)*

La Araucana became a 16th century bestseller and as a result the term *Los Araucanos* spread throughout South America and Europe. The Mapuche are still sometimes referred to by this name. They themselves will say they are Mapuche and they will also point out the poem is wrong — they were never vanquished by the Spanish.

There is a small tributary of the Bio-Bio River in central Chile that is called Arauco. The Mapuche word for this river was first transcribed *rauco* by the Spanish, then later as *arauco*. In Mapudungun: Rag (ra) is the word for clay, Ko, translates as water. To the Mapuche, rauco, misspelled by the Spanish or not, is a body of water, a clay-colored body of water, not a tree and certainly not a people. It describes a certain river, a particular place and this place is not where these trees grow. In his journals Hipólito Ruiz adds his two cents regarding the naming of this species, pointing out that Antoine-Laurent de Jussieu named the genus *Araucaria* from his belief that these trees grow in the area the Spanish called Arauco even though this is coastal territory and these trees grow a great distance from the sea.

This long struggle to assign two meaningful words to this tree has had little impact on the tree itself; it is a label, a simple human conceit. The Mapuche have always called this tree the pewen (pehuén) — they still call it this. They do not refer to themselves as Araucanos, nor do they call the tree, la araucaria. Using these monikers only serves to remind the Mapuche that we really don't know who they are and, seemingly, do not care much about their history.

words to image

Words have power but they have power because they become welded to an idea, affixed to a concept. Sometimes the trick of a word is that the idea or concept associated with it can change from person to person. This is how, so effectively, miscommunications occur personally, politically and certainly culturally. The Mapuche refer to non-Mapuche as winka (spelled huinca in Spanish). It is a word Argentines and Chileans often automatically take to mean *white man* assuming that skin color is the characteristic that distinguishes them the most. But when cleverly using the local lingo to refer to themselves and using the term huinca, what the Mapuche hear them say is, "I am an outsider, a stranger, an enemy." This Mapuche word is not shaded any color, either you are a person to be trusted, or you are not. It is probably safe to say that for the Mapuche and indigenous peoples around the world white quickly did become a color not to trust.

Varying cultural attachments to the same word is not an uncommon occurrence and translation a layered and tricky thing, but words are also influenced by visual attachments. Frequently, our visual sense tags an image to a word and that image often overwhelms the word, controls it, so to speak. It can change our relationship to the reality of a word. Take, for instance, what the term *volcanic spring* brings to mind—that a volcano, of course, looms somewhere in the picture. Change these same exact waters from volcanic to geothermal and place them beneath a building called a spa, a thermal spa and the image changes from potentially explosive, catastrophic activity to one of warmth and relaxation. Same bubbling, possibly volatile water but a big jump in connotation. Both the small villages of Copahue and Caviahue exist because of the thermal waters of Copahue. But Copahue was the name of the volcano before it became the name of the spa. It will be the name of the volcano even if the spa, for some pyroclastic reason, disappears.

Long before the Argentines and the cement buildings showed up, people came to look for a cure in these volcanic springs. A local legend indicates the Mapuche once had a different approach to these thermal waters. To soak in these curative waters it was necessary to first ask permission from the Mapuche chief in whose territory the waters bubbled. Then a request to the *Arünco* or spirit owner of Las Termas must be made. To petition the *Arünco* a person must throw a close personal object into the "eye" of the gushing water. (Alvarez, 1992) If it sinks, favor is granted and the cure needed will be received. If it does not sink the *Arünco* is angry and will not grant what is requested—might as well just pack up your troubles and go home.

Currently, at Las Termas, the procedure is slightly different. A person must fill out a form, consult a doctor, pay some money, then choose the appropriate tonic to drink, waters to soak in or mud to wallow through. Every morning at the hotel as I headed off into the hills the rest of my hotel mates boarded a bus to take them up to the spa. I have a very similar reaction to doctors and potentially angry spirits—unless in serious need, I find it best to avoid either. However, after the perpetual greetings of: *Going to the spa today? Have you been to the spa?* I finally decided to visit Copahue, Las Termas. One overcast morning, I walked down to the *Lavadero* to get a ride up to Copahue with Adrian. He and his wife Debora ran the laundry in town. People often stopped by the laundry just to visit and more than one evening I found myself there drinking maté and folding sheets surrounded by a welcome humidity and the rhythmic hum of the clothes dryers. By the end of my stay, I had been promoted to manning the ominous looking sheet dryer. Feeding sheets into this high temperature, mammoth drum requires quick fingers, teamwork and one's complete attention — all important life skills I discovered I could use some practice in.

After Adrian dropped me off at Las Termas, I poked around the spa while he delivered linens. I skirted the edge of the yellowish sulfur bath, the green algal bath and the brown, appropriately but not so appealingly named, pig bath. I was feeling exceedingly healthy that day and could not think of anything to ask the *Ariinco* to cure. Not one to present myself in false circumstances where spirits are concerned, I decided against a dip in any of the green or milky gray waters, I chose not to wallow in the acidic mud. The laundry run only took a few hours and on the ride home Adrian asked what I thought of Copahue. I hesitated with my reply. There are no trees in Copahue. The village is set down in what looks like a pit of ash; gray steam hisses and billows from the earth and pools of water encased in cement bubble in odd opaque colors. It smells. Dantesque was the first term that came to mind but I did not want to offend Adrian's home turf. Sheepishly, I replied, *Un poco feo. ¿No?* Adrian laughed. *Yes*, he agreed, *Copahue is a little ugly.*

We talked instead about Caviahue, the mountains, the trees. Adrian had been in the Merchant Marines; it is where he and Debora met. They had been to New York City, Seattle, Miami, Tokyo and on and on around the world. How did they end up here, so far from the sea? Quiet living, a remote spot—his smile widened as he said —*you should see Caviahue in winter*. I had heard this before from Ernesto, the same emphasis on winter. The next day, I would hear this again from another equally passionate resident of this small Andean town.

The next day, as I knelt beside Lago Caviahue in a cold gray drizzle, I was thinking about a quiet, simple life. A quiet life even in such a cold place was starting to sound good to me and there is not much I dislike more than being cold. I was carrying a lot of gear with me and Junior, my dear friend, had already knocked me over, twice, into the black, wet sand. Playing tag was on his agenda for the morning; he was not too interested in my taking water samples. To be honest, neither was I that day. On the hillside across from the lake two young boys were launching rocks at the highly positioned cones of a pehuén. In between dodging Junior, keeping my camera and my self out of the lake and filling bottles with acidic water, I watched these two. The desire to trade in my pack full of techno-gadgets for a rock that would reward me with a mother lode of pine nuts seemed a reasonable one. Caviahue, I could feel, was working on me. Was it the stunning quiet or the density of the stars, the imposing mountains or these ancient trees? Was it the warmth and pace of this small town, the contentment in the friendships I had made so easily? Caviahue was no longer such an odd and strange place to me; I was starting to see and feel things as familiar. I had expected to find Copahue, a volcano, a spa, a remote place in the Andes. I had discovered Caviahue, a mountain valley, a quiet town—a tranquil, friendly spot in these beautiful unending mountains.

It is doubtful the town of Caviahue would exist without Copahue and its thermal waters but the draw of this provincial park is more than just the spa, it is also the distinctive ecology of the place and it was in this unusual setting that I was learning about these ancient trees. I had not experienced the pehuén outside of this high altitude valley. There are two other species of tree that grow in Caviahue both of which are southern beech, *Nothofagus pumilio* and *Nothofagus Antarctica* but the combination here of elevation and climate cause these deciduous trees to appear in a 'krummholz' or stunted bush-size form, leaving only the pehuén to tower on the mountain side. Starting at approximately 5,900 feet, a pure stand of pehuén stretches up the mountain side to form the tree line here.

Although these archaic trees help to animate this landscape by bringing the past visually to life, they also exist elsewhere in a variety of different settings. Here they withstand the harshest of elements but the pehuén can also be found at elevations as low as 1,968 feet above sea level. The present small range of this tree also encompasses a wide variation in annual precipitation, anywhere from 47 inches of rain to quite a wet and soggy 156 inches. In Caviahue, the annual precipitation is about 78 inches which falls on these pehuéns mostly as snow. And though these trees are usually associated with the Andes and their volcanic landscape there are two small populations of pehuéns in the Cordillera de Nahuelbuta, a range more toward the coast of Chile where the soil is not volcanic in origin but from ancient granitic and metamorphic rocks.

Currently, these trees are found only in a small area of Argentina and Chile but this small patch was much larger in the not too distant past. According to the *Botany Report of the Princeton University Expeditions of 1896-1899* a hundred years ago the range of this species spread from 30° 30' in northern Patagonia south to 45° 30' in southern Chile. This rapid reduction in range is a familiar story, quickly explained in a few words published in the *Nautical Gazette,* London, 1919, *the wood (of the pehuén) could not be better suited for shipbuilding.....* It is hard to guess what the range of the pehuén was prior to this *Princeton Report* but in 1781 when Colonel Higgins sent troops off into the hills to look for a replacement mast for the *San Pedro de Alcántara,* they found within *thirty leagues of Santa Juana* at least seventy pehuéns of the right size and height to replace the mast. They also reported seeing 300 more trees, 50 to 80 feet tall, between Angol and the coast. Geographically, this is the same area of the two small populations of the Cordillera de Nahuelbuta, the only surviving stands of pehuén outside the Andean Range. Hipólito Ruiz, in his journal, reports that: *According to the natives, the forests of this species of pine stretch over 200 leagues; that is, from 36° almost to the Strait of Magellan.* (Ruiz, 1940 ed.)

I had come to the Andes toting the words old, ancient, archaic, to look at a tree that now only grows in a small range of Argentina and Chile. A relict species, the ancestors of which were widely found across both the Northern and Southern hemispheres during the Mesozoic Era. This is what I knew of these trees and when I arrived I found that my idea of an *archaic tree* was startlingly reflected in the appearance of this volcanic valley. Experiencing these trees, in this landscape, only made the association to their past more striking, more direct. But something was now bothering me about these trees and their Jurassic juxtaposition.

juxtaposition

Naturalists by occupation, or preoccupation, identify things. And when they are uncertain as to *what* something is, they know the importance of seeing *how* it is. Visual artists, though not usually concerned with this naming of things, are also intent on understanding how a thing appears. To interpret what they see, they need to visually comprehend its form. I have learned from spending an inordinate amount of time with both these types of observers that the naturalist and the artist share a way of seeing the world. Both use contrast, scale, shape, color, and texture as tools of comparison, as meters of understanding.

This dynamic of visual comparison enhances memory and appreciation for the naturalist and for the artist inhabits photographs, paintings, poems, any place where the visual elements seep together, form and reform into thought. It is here in this shared and heightened sense of noticing that an exchange of information can be invaluable, that the artist's way of seeing and the naturalist's perception of things can enhance one for the other. The place to which the artist and the naturalist takes this visual information may be different, but the process, how we teach ourselves to see, is much the same.

As children, we are taught mostly through memorization and association and are not given much in the way of perceptual skills. So when we are presented with the word *tree* to describe say, 625 North American species of woody perennial plants, we generally leave it at that and move on to other things. We are taught the color of the sky is blue even though on any given day that sky can surround us in enumerable blends and hues of blue and orange and gray. We do not question the vagueness of these words until something causes us to perceive them differently, to examine, to compare, to look past the rudimentary words we have been given. It is then, when we start to take a closer look at our surroundings that we find that our perception of *sky-blue* is faulty, our sense of *tree* severely lacking.

The reasons a young artist or a budding naturalist examines their environs more closely are different, but the sense of awe that unfolds as they start to perceive a vast array of new details is much the same. There is a defining moment when a person perceives that the blue sky that crowns their everyday life cannot really be successfully reproduced or summarily expressed by that crayon or tube labeled Sky Blue. This realization is much the same as the moment another person looks out and sees that the homogeneous swath of green formerly known as trees, forest or woodland is actually a wide and varied community of plants, hundreds of them in fact. Instead of seeing one green mass, distinctive shapes and textures emerge to forever change the identity of that same landscape. It is a startling and exhilarating moment acquiring that new sight. Learning to pick out the silhouette of the oval crown of a sugar maple from the asymmetrical flounce of a red maple is much the same as learning how to see, mix then lay down the right yellow-green that will express the airy foliage of a weeping willow. In both cases the nature of each tree, its shape, texture, line, pattern, hue, is considered, its individual identity seen, compared and remembered through juxtaposition.

Our culture has developed a strong penchant for specialization and both the fields of natural science and the visual arts have their own languages in which to converse. Ordinarily, this makes it more difficult to use each other's tools, but in the translation of terms dealing with the process of observation, of seeing, these two disciplines can, in fact, understand each other perfectly. Naturalists might not speak of juxtaposition but they understand in a moment that it is a concept they use consistently out in the field trying to discern one species from another. The artist might not choose to elaborate on the nature of a shape as ovate or imbricate but the conveying of these bits of information might provide an important visual concept that enable them to bring a certain understanding to their work. However, it is not really important whether these shared aids to perception have specific applications or not. New ways of seeing are invaluable. To be locked into our own habitual pattern of observation limits what we can know; if something is seen in only one manner, in one light, our understanding of it can never expand or evolve. Juxtaposition, the examining of one thing in relation to another, makes us think, not only about what we see, but what we know about what we see.

The gray day, the white-capped lake, everything with the exception of one hyperactive, red dog had put me in a reflective mood that afternoon by the lake. That night the sky was black. It was not the transparent black of a deep and endless mountain view out into our Universe. There was no indigo horizon darkening gradually up behind the stars. It was a thick, low hanging black, obscuring the distant mountains and cropping off the tops of those closer by. There were no silhouettes, no staggered outline of pehuéns spread across the ridge with their amazingly long, straight lines draped at the top with a scribble of branches. Even the volcano had vanished. I sat in my room and watched as lightening and wind from an approaching storm headed into the valley. I thought about this valley as it was rapidly disappearing behind the low gathering clouds. How easy it was here to be deceived by familiar forms. The valley itself is not truly a valley but the floor of an ancient caldera, the beautiful, blue lake is full of water but the water itself is empty. The mountain behind the hotel has a small ski slope at its base. It is a gently sloping, snow capped and seemingly quiet place but, in truth, it is a place of violent reactions – an active volcano. The town is young, the trees are old, but just how old and tall it is impossible to judge. From a distance their tallness is lost, no other trees grow up beside them, there is nothing by which to compare their looming presence. Had I just passed through this town, Lago Caviahue would have appeared to be a normal lake, the valley just a valley, the walls of the extinct volcano just part of another range of mountains and the active volcano simply a remote but safe place to swoosh downhill through deep, powdery snow. The juxtaposition here of identity and reality had provided an amazing place to hone my observation skills. Here, I had learned, it was very wise not to take *anything* for granted.

The storm finally arrived bringing booming mountain thunder echoing in the darkness and lightning so close its electric charge kept me from my bed. I waited for the arrival of the rain to try for sleep but it did not come. I fidgeted with the equipment, my camera, whatever I could find to distract myself. I picked up a notebook feeling the need to organize my thoughts, my life, at least the next few days. My stay was coming to an end and I needed a plan for obtaining the last two water samples. One from the far side of Lago Caviahue where the Rio Agrio runs out the other side of the lake and down the valley and the other from the crater lake. Sea level storms leave no portent with me, they clear the air or bring a magnificent strength to the sea; they are like moods that come and go. But being so high up, so close to these jagged bolts of lightning and not sure what to expect next, I felt the need to pay attention. I stayed awake but accomplished little.

Finally, I slid between the white and softly worn sheets of my bed. I watched each rent in the blackness, every jagged rip of light, flash and disappear, flash and thunder and contemplated this seemingly harsh, high altitude landscape as it appeared and disappeared. My sense of this valley, these mountains, its trees, had shifted. Nothing was elementally different, but I lay in the dark that night reworking the picture. Something had changed my perspective. I tried to place some image, a sound, a shape, one word, one moment next to another trying to figure out what made this landscape now, for me, even surrounded by a violent mountain storm, feel so absolutely hospitable. Even though, in this place, the earth was cut with acidic water and formed by flowing rock instead of the time-nurtured soil with which I am so familiar. I imagined I could not sleep thinking about this rearrangement of things, but when the lightning finally moved on, I drifted off. Somehow I had forgotten to worry about my climb up the volcano.

The next morning the sky was lighter gray; the clouds still hung close. I walked through the quiet, *still* dusty, town. It had barely rained, hardly a drop for all that noise. Outside a wooden A-frame I found Ernesto's friend Rubén under the hood of a Willys jeep. The day before I had stood in this house drinking maté with Analia and Ruben's wife, Patricia, examining all the rocks and fossils that lined the window sills, noticing the cross-country skis, the snow shoes, the rappelling gear stashed in the eaves and hanging on the walls. Ernesto had showed up later and then Rubén arrived, carrying kayak paddles in his hand, his wetsuit still dripping water. I looked out to the lake, it was the only place he could have been but I had never seen anything or anyone out on that acidic lake. Before Rubén stopped dripping, Ernesto started making arrangements for Rubén to take me up the volcano. I stood there mutely, being discussed. Ernesto pointed to my heavy hiking boots and assured him I was prepared for such a trek. Truthfully, I don't think it was my footwear Rubén was concerned about but he finally agreed to take me up to the crater lake and after he changed into dry clothes plans were made over more maté. They decided we would go the day after next; the heavy clouds should lift and, hopefully, it would not snow on the volcano before then. The locals called Rubén, Caniche, or *poodle* as it translates, for his long, dark ringlets of hair. Language is an interesting thing; had the word poodle not come up, I would have also used this nickname, Caniche, it has a nice ring to it. But I could not get by that word, poodle. They are smart dogs I know, but people put bows in their hair and make them do stupid tricks. I preferred to call him Rubén.

The Willys jeep never started. Instead, Rubén gave me a lift to the other side of the lake in a very noisy, yellow truck, a snow vehicle of some sort, ten gears — all low. The truck vapor-locked three times. By the time we got to our destination Rubén was smeared with grease and had ingested a certain amount of gas but not once had he cursed (at least audibly) or kicked at the difficult truck. The sky looked like it might clear. A harrier or two flew low above the dry grass of the valley. Rubén grinned when I asked my usual question—*yes, if luck was with us, we would see a condor, maybe two.*

After getting the water sample we drove along the river, out through two red and crumbling mountains to a place called Salto de Agrio. Here the Rio Agrio flows off the high mountain valley floor, over a basalt cavern and drops deep, spreads wide into a dark-blue pool. The water thunders down this high fall to meet its own splash and spray on the way back up. The gray clouds had finally blown off and white ones moved fast overhead as the sun gradually took back the sky. Grinning, Rubén told me how he rappels down alongside these falls, how hanging there, very still, next to the moving water makes it feel like he is falling. I paid close attention to the fact that he said this as though it was a *good* thing.

We talked about the surrounding mountains, about backpacking; he pointed out the mountain peaks he had scaled or the faces he had rappelled. I questioned how well this rock held; not as good as yours, was his reply. I wasn't sure I had understood him correctly. New England's mostly granite mountains are made of harder stuff but did he know that? Did he understand about the shady Green Mountains I hike in, how they are different then the Tetons, the Rockies, the Cascades? I wondered just how specific a conversation were we having about the mountains of North America. Between stories of the winter in Caviahue, the trees, the mountains, the pumas, he told me of his travels unwinding through most of South America, Spain, France, Italy, Morocco, Senegal, Germany and then there was the repelling, the kayaking, the white-water rafting and a marathon in Hawaii. By the end of the afternoon I knew there had not been a miscommunication. He knew the topography of New England, and he knew exactly how I felt about both my Green and White Mountains.

These mountains of his were red; pumas still hid in them and condors flew over their peaks, at least they did that afternoon. Rubén's smile changed magnitude as he pointed up into the sky. It was thrilling; they were high and soaring but juxtaposed against the tops of these mountains I understood the huge breadth of their wingspan. Black against deep blue they wheeled in circles, flat winged and gliding until finally, with barely a flap, they disappeared up and over the peak. We both stared up at the sky until only blue remained between the clouds. The sun was warm, a breeze blew a mist off the thundering, falling water and luck had, that morning, just been with us.

On the drive home we broke down only once. Rubén explained the winter, the snow, and the need for this truck. Copahue shuts down for the season and Caviahue shrinks to about 300 people; no one goes to the Spa, you can't get there. Lago Caviahue never freezes but if it were the kind of lake that could, it would be buried under the six or sometimes nine feet of the snow that falls here in the winter. The road is closed to Copahue and it's difficult to get anywhere, unless of course, you have a yellow snow truck, cross country skis, snowshoes or a team of huskies. Rubén pointed at a house back off the road. *There,* he said, *they have a team of eight.* I knew by the look on his face what he was going to tell me next. *The winter, it is beautiful here.*

It was still early in the day so after we parked the truck back at the house we walked out of town, up into the valley just below the volcano. We hiked around the escarpment that held Lago Escondido above us. It seemed a random hike but a few hours later when we stood beneath one of the grandest trees I'd ever seen, I realized there had been a destination all along. The tree, dwarfing all the very tall pehuéns around it, had to be at least a thousand or so years old. I took a photograph. It is not a picture of Rubén beside the tree, it is impossible to stand *beside* anything this tall. It is not a photo of Rubén himself, he is discernible only on second glance, and it is really not a representation of the tree — much of the top is cropped out. This photograph with its juxtaposition of man and tree is an image about time. Within that frame, that image of *Time* in the context of this place and these pehuéns, finally made me see what had changed my sense of this place. It was, of course, these tall, long-living trees. They had stood a very long time in this place — this volcanic mountain valley was not an easy place to stand.

context

Perspective, scale, proportion, point-of-view and juxta-position—these things all inform our perception, they help us to focus, but once we "see," where do we place this sight. Context is how we choose to frame what we see. In viewing an unfamiliar work of art we might immediately place this visual experience within our own context. It is in this framework that our intial response is formed. But if we choose to look at the work in the context of the artist, or that artist's place in history, it can shift how those same elements are perceived. Though in the end it might change us, this new context does not change the piece itself. It is still there to return to, to look at, to see. Consideration should be given to both these ways of experiencing the work. In observing nature, as in art, there is no difference. Recognizing the value of our own visual perception and response is important as long as we understand the context is ours. In the same respect, dismissing those perceptions because they do not fit in a context one has come to "know" shuts down the process of discovery and limits understanding.

An active volcano provides a good deal of context for those living below it and this volcano, Copahue, provided the context of my presence in this out-of-the-way place. But like many of the residents in this town who don't talk much about its potentially explosive presence, I too had been giving the volcano a cold shoulder. I am already awake when my alarm goes off and I lie in bed thinking about climbing the volcano, meeting my first bubbling crater lake and procuring some of its acid water. I am excited about the hike, the high altitude climb and a day spent in good company. I am curious to finally peer down on this lake with its ominous sounding cloud of hydrogen chloride, a cloud I had been advised to avoid. I try to keep any pyroclastic image from my mind while at the same time accepting the context of this climb. Ascending a volcano is much like taking off in an airplane, once up there things are pretty well out of one's control. It is still dark and I force myself to get up. I have just two more mornings left, two dawns to watch, two more days to remain in this place and too much to think about this early in the morning. I need coffee.

I think about Junior as I get out of bed. Our paths had not crossed these last few days and they would probably not do so again today. An image of Junior's exuberant self runs through my mind. The image of Junior accompanying me up the volcano, of Junior bounding along the icy rim of the crater and of Junior wagging his beautiful, red tail as he knocks me into that boiling lake. It seems a much better plan to be hiking with Rubén. I find my gloves, goggles, hat, pack my meters, notebook and camera, rig-up a long rope to a gallon bottle before I pull on my long underwear, heavy socks, boots, and then slip, well, actually struggle, into my Gore-tex jacket.

The wind that swooped over the mountains during the night is still blowing strong as I step out into what isn't quite yet a new day. Though I do not speak it all that well, Spanish has given me more words with which to think, some of these words handle concepts that one word of English doesn't quite manage. But the time between night and any early morning moment is not separated any further in Spanish than it is in English. Pre-dawn is not a very poetic word and the word twilight, though it means *the time when light from the sun is below the horizon*, seems to have been appropriated by the end of the day.

Because there are no other words in English to take us from night into day, pre-dawn and post-dawn must suffice. In Mapudungun, however, if I trust in translation, these moments have their own words. Dawn is *aùn*, that instant the sun begins to illuminate the countryside. *Pichiantü* is a time of the first light of day, or the time a day is still new. *Lihuen* is morning and *huelihuen* is before the sun has risen. But *Allafün* is the word I have been looking for. A word for when dark shifts to deep gray, a time when that first bird trills one solitary note. *Alla* means beautiful, pretty; *fün* signifies fruit, ripening. *Allafün* is a *beautiful ripening* — morning twilight. The Spanish translation by Erize Estaban in his Mapuche/Spanish dictionary explains this as: the *beauty of the instant which the stars catch sight in this subdued light of the rapidly emerging scene that surrounds them*. It is evident Estaban has taken some poetic license with this translation but I am enthralled by it all the same.

Allafün. One poetic word for the soft crossing of night into day, a time neither night nor day can claim. A moment gray gets to be an uplifting color, to be anything but a somber, dreary thing. I walk out into this almost gray and down to a small stand of pehuéns on the outskirts of town. I stand among the trees braced against a persistent wind, taking in the shifts of a sunless and still shadow-less landscape, taking in the *allafün*. The stars start to slip off and this earth, those trees, these mountains will soon be visible for another brief day. Every shade of gray imaginable is before me and then, as I watch the gray disappear like smoke, the colors of the earth start to slowly appear. Silhouette after silhouette of each grand pehuén emerges then disappears, as dark shape becomes green tree. I realize watching this odd, but now familiar, landscape come to light, that through the course of all these mornings sitting, watching it dawn around these trees, how my perspective, my point of view, has changed. I look out over the valley, the volcanics, these same trees, but now I no longer see the past.

I came here to learn about these trees and my attachment has grown deep, but my perspective of them has changed. I see them now in the present, for me they no longer simply stand as symbols of the past. I crouch down between the rocks below the pehuéns trying to get out of the wind and I sit and watch another dawn. Surprisingly, climbing Volcán Copahue isn't on my mind; leaving Caviahue is. The quiet here elicits respect, respect that a peaceful place deserves and I realize that for some, it is not Las Termas, nor the volcano that is the context of this small mountain town, it is this quiet. It is this peace that I do not want to leave. I have come to understand a great deal about this place but everyday I have learned how much there is I still don't know. Nature is like that, one endless lesson. But today's lesson is not about trees, it is about altitude and oxygen and volcanoes and it is time to go meet Rubén. I walk back into town.

Ernesto and Analia decide to go with us and driving up to Copahue the wind starts to push Ernesto's little blue Honda from the dirt road. We park across from Las Maquinas, a tiny hydro-thermal plant, and sit looking at the plume of very forceful steam bent by the wind and streaming almost at a right angle to the vent! It is very windy. Once out of the car we all try to manage our whipping hoods, strings and zippers to suit up against the intense wind. Now I decide to start worrying. We have not even started to climb and the wind is furious.

Words blow by too quickly to hear, eliminating any casual conversation as we start the trek. But even without the damper of wind, thin mountain trails, whether they snake through a dense wood or follow a high barren ridge, mandate a quiet companionship, a singular walk. We fall into the rhythm of our own thoughts. My thoughts are on the wind. Several times I lose my balance as a gust hits me broadside and in mid-step. Analia, even smaller and lighter than I, gets knocked over several times before we both figure out how to manage its push. We walk for several hours, stopping only occasionally to catch deeper breaths in the thinning air. Ahead of me Rubén's steady, slow pace reassures me he has walked this high altitude walk before.

El soroche is what it is called if you are standing on the Chilean flank of one of these mountains but here in Argentina the affliction is called la puna. On either side of the Andes it feels and translates the same — altitude sickness, when the lungs cannot extract enough oxygen from the air to pass into the blood stream. It is something to pay attention to; a painful lesson I had learned once already in another hemisphere. My sea level lungs are keeping up with the task and I am managing to keep up with my companions whose hearts and lungs are always in these mountains but I begin to realize that the thin air is the *only* thing I do know to expect from this place. It is a foolish time to begin this thought process, a pointless moment to consider the wisdom of climbing around on this active volcano. It is a useless time to wonder whether I should have tried harder to find that gas mask. But I do all this anyway. I have the silence imposed by a loud wind and a long, quiet climb to mull things over. I begin a solemn review of my life's many misjudgments.

Analia gets fatigued and she and Ernesto drop behind. Rubén starts to turn around more and more to see if I am still with him. We keep walking. When we hit the ice pack Rubén stops to look behind. Ernesto and Analia are far below and Rubén, looking to the west at a bank of dark gray clouds, decides we should not wait for the two of them to catch up. He points ahead, seemingly straight up, to a craggy v-shaped gap in the rim of the crater. "We are close," he says. The west wind is pushing a steady stream of clouds, vapor, something—straight out of this gap. The thin line of a trail leads to the bottom of that cleft in the rim and I start to have an uneasy feeling about the origins of that cloud. I pull off my pack and dig around until I find the goggles and the two, probably uselessly silly, paper masks I have brought. I offer a mask to Rubén; he takes it, puts it on and I follow him into the cleft.

With one whiff I realize that the cloud is not blowing in from over the top of this mountain, it is rolling off the surface of that roiling, acid lake. I also realize I never bothered to find out exactly what effects a hydrogen chloride cloud might have on one's lungs. Rubén does not seem too concerned but when I lose sight of him in the cloud, panic brings me to a halt. I am not a brave person. What in hell am I doing here? I think of what I have come to know about Rubén; he kayaks in acidic waters, he rappels down the face of crumbling cliffs. I flash on the story of him letting someone, somewhere in the wilds of Brazil cut him open—it was not, I can assure you, a doctor in the hospital sense of the word. I had seen the six-inch scar across his shoulder blade, inch-size Frankenstein stitches and all. This was not an overly cautious man and here I am following him into this gaseous cloud toward a crater lake and I cannot see more than a few feet in front of me. "Rubén!" I call his name. "What's the matter?" He is back at my side asking. "I can't see!" is all I manage to explain. "You need the water, no? We are here." Rubén points down. I look to where he is pointing and there it is—the crater lake. It is wild, the wind is whipping across its surface and my mind is a complete blank. "The bottle," he says to me, "give me the bottle." I take off my pack and pull out the bottle and the thick loop of rope. He takes it, lies down and leans out over the edge as I hold onto his legs; the spray from the lake splashes us. It takes only a second to register the alarming discrepancy—he only needs about six feet of the fifty-foot rope I was told I would need to drop from the *rim* of the crater. We are definitely not up on the rim of the crater. Rubén hauls up the water sample and then we get the hell out. It is not until we stop about ten minutes down, and I find that our lungs still work and our eyes, though bright red, still see, that my brain begins to function once again. It is then I remember—I had forgotten to take the temperature of the lake.

Ernesto and Analia wait for us down below, near a fissure in the volcano's side which I realize is the headwaters of the Rio Agrio. When we reach the two of them we stop, throw our packs down and collapse, leaning back into the steep slope. Rubén's jacket, I notice, is dotted with tiny acid holes. We sit far enough away from this toxic stream as not to inhale its sulfuric fumes and savor our meager but exquisite lunch of ripe pears and dark chocolate. Ernesto and Rubén discuss the crater lake. Ernesto wants to hear every detail; particularly he wants to know the temperature of the lake. Unfortunately, I have to tell him—I don't have a clue. Temperature fluctuation is important in monitoring a volcano's activity and Ernesto, after all, lives below this volcano.

I lean back on a large chunk of relatively newly made rock and stare out over the distant valley; the drama, my fears, my failing all slowly drift away. For the first time, out of the wind and away from the inside of that crater, I take in this amazing place. A glacier, no longer melting under a summer sun, stretches out below us, dusted with ash. A cleft to the right of where we sit exposes strata of purple and dusty gold stone swirled together forming a dramatic wall that looks more river then rock. Large boulders are strewn everywhere and all around us the newly flowed and spewed look of the ground on which we sit speaks of forces hard to fathom. It is quiet here, at this moment, but it does not resonate as a quiet, serene place. Sitting at the top of an active volcano is too literal a place—it is impossible to ignore all the visual evidence of its nature—quite clearly this is a violently made mountain.

The context in which I came to know *Araucaria araucana* is that of these mountains. It is here I came to understand the height and breadth of these trees. It is here I saw how tall, how wide, a thousand years growth can be. As the west wind continues to blow off this volcano and down into the valley, I think about all this. I think about the nature of these trees, how they grow tall and old in the harshest of elements. I now find it difficult to think of this tree as anything but a hardy species trying to hold its ground, placed in a precarious position by man. To see this tree in the light of another time, a time when the family *Araucariaceae* included numerous species and was found in both the Northern and Southern Hemisphere, is to place this modern-day species in the context of its ancient ancestors. *Araucaria araucana* is one of the few extant members of this family and this attaches the words *relict species* to this tree but these words, so attached, can bring with them the idea that, from a natural selection point of view, it has not been successful in its modern habitat.

What also influences this view is that in the *Araucariaceae* family's heyday conifers were far more widespread than they are today. The increase of angiosperms over gymnosperms since the late Cretaceous Period gives a competitive spin to the relationship of these two groups of trees. Studies in the past have been done looking at the small range of this species as being a result of its inability to compete with the broad-leaf inhabitants in the same areas. But studies of the regeneration patterns and succession of *Araucaria araucana* in the forests of Chile and Argentina (Veblen, 1982, 1977) (Burns, 1993) indicate that this species is successfully reproducing in its range and is not being replaced by the angiosperms that share its habitat. They also indicate that *Araucaria araucana* is better adapted to a more stressful environment than the deciduous trees that also grow in these mountains. A genetic study (Rafii, Dodd, 1998) done on three Andean populations and the isolated coastal populations of *Araucaria araucana* supports the perspective that these populations of *Araucaria araucana* should continue to remain viable, withstanding, of course, any human influence. This study, looking at the proportional composition of hydrocarbons in the waxy coating on the leaf surface of *Araucaria araucana,* indicates that this species is characterized by a relatively high degree of within-population genetic variation. This variation supports the idea that there is genetic adaptation to the more arid environments that in turn suggests the viability of this species in the face of moderate levels of environmental change.

The Andes Mountains have a way of rearranging a landscape. Volcanic eruptions, landslides, mudflows, these things do happen here and conifers appear to be better at maintaining in, and recolonizing after, these massive disturbances. Caviahue, where the pehuén dominates, is not a place that deciduous trees grow well but some see the pehuén's success in this harsh spot as simply relegation to an unfavorable site and therefore evidence of its relict status. Because man has so reduced the range of the pehuén it is certainly possible that further small population extinctions could bring about the end of this species. But a population extinction is not the same thing as a species extinction, even if the one brings about the other. To see this species as a "relic," to frame it in a geological time-scale of millions of years, is to set the stage for the future of this "un-evolved" species in extinction. To designate a species, even a protected one, as headed toward a natural end of its time on this planet is a dangerous thing. The words *natural end*, like the phrase *died a natural death*, are used to assure someone that the death was inevitable, no one is to blame, nothing could be done. Goethe points out in his *Scientific Studies*: *How difficult it is to refrain from replacing the thing with its sign, to keep the object alive before us instead of killing it with the word.*

Images and words are powerful things but nature always finds a way not to be denied its strength. The view from the top of a mountain is something an image cannot replace. No photograph ever quite captures it or creates an expectation of beauty that the mountain can't deliver. A painting or drawing may touch on the essence of this spot but viewing this work will not flood the senses in this same high altitude kind of way. I stand in this thin air, on red and black chunks of raw earth, looking out at the drift of distant, blue mountains wondering how rapidly perspective would shift here in the first light of day. But allafün, the morning's twilight, dawn, is a night away and a cold, dark volcano is not the place I care to sit and wait.

I will spend my last time in these mountains waiting for the stars to catch sight of this young earth leaning back against the reassuring girth of some great old tree. A tree that stands sentinel in these mountains, this valley and looms over a forest of other giant trees. Trees I have come to see in a different light, trees that have managed life on this earth for a very long time, trees that express stability on this sometime shaky ground. I do not believe there is just one fixed way to survive on this planet. In nature, there are species that work well in their own particular niche and do not need to evolve into something else and there are others that have to transform very rapidly to maintain a place for themselves. The evolution of our culture into a highly technological society, an ever-increasingly mono-culture of bigger, better, faster could be the evolutionary edge we assume it to be or it just might be our demise. As a species that is so self-assured in our destiny, I think we purposely overlook what nature so prolifically shows us — there is no singular path to survival. Instead, our aggressive, fast moving culture believes that there is only one path to take and it is ours. The rest of the planet needs only to follow in our very busy footsteps and they too will survive. If these trees have shown me evidence of one thing, it is that the strong do survive — but that strength does not always manifest in one guise and change is not always the key to survival.

As Ernesto and Rubén head off down the trail together they sing out to the valley below a castellano tale, a melodic folk song that they both know by heart. I gather my thoughts and my pack and head down after them. Analia stops and waits for me to catch up. The sky is cloudy but the wind has died down and it pushes only gently at our backs.

bibliography

Aagesen, David L. (1998) Indigenous resource rights and conservation of the monkey-puzzle tree (*Araucaria araucana*, Araucariaceae): A case study from southern Chile. *Economic Botany*, 52 (2) (146-160). The New York Botanical Garden Press, New York.

Almeida, Aimara Riva de (1979) *Bibliography of the Genus Araucaria*. Fundação de Florestaís do Paraná, Curitiba.

Alvarez, Gregorio (1992 ed.) *El Tronco de Oro*. Siringa Libros, Neuquen, Argentina.

Armesto, J.J., Villagran, C., Aravena, J.C., Pérez, C., Smith-Ramirez, C., Cortés, M., and Hedin, M., (1995) Conifer Forests of the Chilean Range. In Neal J. Enright & Robert S. Hill, (eds) *Ecology of Southern Conifers*. Smithsonian Institution Press, Washington, D.C.

Bangert, William V. (1986) *A History of the Jesuits*. Institute of Jesuit Sources, St Louis, MO.

Beardmore, John A. (1983) Extinction, Survival, and Genetic Variation. In Schonewald-Cox, et al. (editors) *Genetics and Conservation* (125-151), The Benjamin-Cummings Publishing Co., Menlo Park, CA.

Beloff, Max (1962) *The Age of Absolutism: 1660-1815*. Harper and Brothers, New York.

Burns, Bruce R. (1993) Fire-induced dynamics of *Araucaria araucana-Nothofagus antarctica* forest in the southern Andes. *Journal of Biogeography* 20, (669-685). John Wiley & Sons Ltd., Chichester, UK.

Emerson, Ralph Waldo (1990 ed.) Essays: First and Second Series, Vintage Books, New York.

Cushner, N.P. (1980) *Lords of the Land, Sugar, Wine and Jesuit Estates of Coastal Peru.* State University of New York Press, New York.

Dallimore, W. & Jackson, A.B. (1948) *A Handbook of Coniferae.* Edward Arnold & Co., London.

Darwin, Charles (1969 ed.) *The Voyage of the Beagle.* P.F. Collier & Sons Co., New York.

Erize, Esteban (1987) *Mapuche.* Vols.1-5 Editorial Yepun, Buenos Aires.

Faron, Louis (1968) *The Mapuche Indians of Chile.* Waveland Press, Inc., Prospect Heights, IL.

Farrell, Brian D. (1998) 'Inordinate Fondness' Explained: Why are There So Many Beetles? *Science 24*, vol. 281 (555-559). AAAS Press, Washington D.C.

Grosfeld, Javier y Barthelemy, Daniel (1995) *Arquitectura y Secuencia de Desarrollo de Araucaria araucana en los Bosques Nativos de Argentina.* - IV Jornadas Forestales Patagónicas. San Martín de los Andes.

Harrington, Karl P. (1969) *Richard Alsop; 'A Hartford Wit.'* Wesleyan University Press, Middletown, CT.

Herman, Scott W. (1998) *The Hydrogeochemistry of Copahue Volcano, Argentina.* BA Thesis, Wesleyan University, Middletown, CT.

Houghton, P.J. & Manby, J. (1985) Medicinal Plants of the Mapuche. *Journal of Ethnopharmacology*, 13, (89-103). ISE, Elsevier, Copenhagen.

Hughes, D. Wyn (1982) What Price A Puzzle. *Amateur Gardening*, 3 July 1982.

Koch, Karl (1873) *Dendrologie*. Verlag von Ferdinand Enke, Erlangen, Germany.

Leitch, Barbara (1979) *Indian Tribes of North America,* Reference Publications, Algonac, Michigan.

Lyell, Sir Charles (1877 ed.) *Principles of Geology Vol. II.*, D. Appleton & Co., New York.

Maletti, Ernesto J. (1997) *Caracterización; Área Natural Protegida Parque Provincial Copahue Caviahue,* Neuquen.

Maletti, Ernesto J. (1997) *Pehuén, Araucaria araucana (Mol.) Koch;* Compilación. Administración Parque Provincial Copahue Caviahue, Neuquen.

Mitchell, David J. (1980) *The Jesuits: a History*. Macdonald, London

Molina, J. Ignacio de (1782 ed.) *Saggio Sulla Storia Naturale del Chili.* Stamperia di S. Tommaso d'Aquino, Bologna.

Molina, J. Ignacio de (1808 ed.) *The Geographical, Natural and Civil History of Chili* Vol. I, R. Riley, Middletown, CT.

Molina, J. Ignacio de (1809 ed.) *The Geographical, Natural and Civil History of Chili* Longman, Hurst, Rees and Orme, London.

Molina, J. Ignacio de (1810 ed) *Saggio Sulla Storia Naturale del Chili*. Montessus de Ballore, Bologna.

Morner, Magnus ed. (1965) *The expulsion of the Jesuits from Latin America*. Knopf, New York.

Rafii, Zara A., & Dodd, Richard S. (1998) Genetic diversity among coastal and Andean natural populations of *Araucaria araucana*. *Biochemical Systematics and Ecology* 26 (441-451). Pergamon Press, Oxford.

Rudé, George (1964) *Revolutionary Europe: 1783-1815*. Harper and Brothers, New York.

Ruiz López, Hipolito, (1952 ed.) *Relacion Historia del Viage a los Reynos del Perú y Chile*. Transcribed by Dr. Jaime Jaramillo-Arango, Real Academia de Ciencas Exactas, Fisicas y Naturales, Madrid.

Ruiz López, Hipolito, (1940 ed.) Travels of Ruiz, Pavón, and Dombey in Peru and Chile (1777-1799) translated by B. E. Dahlgren, Field Museum of Natural History, Chicago.

Record, Samuel J, & Hess, Robert W. (1943) *Timbers of the New World*. Yale University Press, New Haven, CT.

Saver, J. D. (1976) Changing Perception and Exploration of the New World Plants in Europe 1492-1800. In F. Chiapelli (editor) *First Images of America Vol. II.* (813-832) University of California Press, Berkeley, CA.

Schultes, R. E., & Jaramillo-Arango, M. (1998) *The Journals of Hipólito Ruiz; Spanish Botanist in Peru and Chile 1777-1788*. Timber Press, Portland Oregon.

Seamon, David & Zajonc, Arthur (1998) *Goethe's Way of Science: A Phenomenology of Nature*. State University of New York Press, Albany, NY.

Soule, Michael E. (1983) What Do We Really Know About Extinction? In Schonewald-Cox, et al. (editors) *Genetics and Conservation.* (125-151). The Benjamin-Cummings Publishing Co. Menlo Park, CA.

Stafleu, Frans A. (1971) *Linnaeus and the Linnaeans.* N.V.A. Oosthoek's Uitgeversmaatschappij, Utrecht

Stafleu, Frans A. & Cowan, Richard S. (1981) *Taxonomic Literature* Vol. III, Bohn, Scheltema & Holkema, Utrecht.

Steele, Robert A. (1964) *Flowers for the King: The Expedition of Ruiz and Pavon and the Flora of Peru.* Duke University Press, Durham, NC.

Swanton, John R. (1952) *The Indian Tribes of North America.* Smithsonian Institute Press, Washington D.C.

Varekamp, J., Delpino, D., Bermudez, A., Deboer, J., Herman, S., and Garrison, N. (1997) The Magmato-Hydrothermal System of Copahue Volcano, Argentina. *Newsletter of the IAVCEI Commission on Volcanic Lakes.* 10 (28-29). Université Libre de Bruxelles, Brussels.

Veblen, Thomas T. (1977) Plant sucession in a timberline depressed by volcanism in south-central Chile. *Journal of Biogeography* 4 (275-294). John Wiley & Sons Ltd., Chichester, UK.

Veblen, Thomas T. (1982) Regeneration Patterns in *Araucaria araucana* forests in Chile. *Journal of Biogeography* 9 (11-28). John Wiley & Sons Ltd., Chichester, UK.

Veblen, T. T., Burns, B. R., Kitzberger, T., Lara, A. and Villalba, R. (1995) The Ecology of the Conifers of Southern South America. In Neal J. Enright & Robert S. Hill, (editors) *Ecology of Southern Conifers.* Smithsonian Institution Press, Washington, D.C.

Wieland, G.R. (1935) *The Cerro Cuadrado Petrified Forest.*
Carnegie Institution of Washington, Washington D.C.

White, Mary E. (1990) *The Flowering of Gondwana.* Princetown
University Press, Princetown, NJ.

Yoon, Carol Kaesuk (1998) A Taste For Flowers Helped Beetles
Conquer the World. *Sciences Times*, July 28, 1998, The New
York Times, New York.

Image notes:

p. 8 - 9
1794 Samuel Dunn Wall Map of the World in Hemispheres,
Kitchin's General Atlas by Thomas Kitchin. Laurie & Whittle,
London, 1797. Provided to Wikimedia Commons by
Geographicus Rare Antique Maps.

p. 58
Illustrations by Davi Leventhal

p. 75
Flore des serres et des jardins de l'Europe by Charles Lemaire and
others. Gand [Gent], Louis van Houtte, 1862-1865, volume 15,
plate 1577-1580.

p. 76
Lithograph from *History of the Indian Tribes of North America* by
Thomas McKenney and James Hall. This lithograph is from
the portrait painted by Charles Bird King in 1828.

Back Cover:
The Lawrence Durrell quote is reproduced with permission of
Curtis Brown Group Ltd, London on behalf of the Beneficiaries
of the Estate of Lawrence Durrell. **Copyright © Lawrence
Durell, 1958**

acknowledgements

To examine any one thing in nature it is important to acknowledge all that surrounds and supports its existence, its growth. With a project like this the same is true. I must begin by thanking my family and friends for their love, patience and consideration. My sister, Susanna Coffey, could not have supplied me with more support and inspiration for this work. The memories of my parents, Edwin and Magel, are woven tightly into these pages in countless ways, their voices resonating at every difficult turn to keep me going forward.

The germination of this book took place in the fertile and unending enthusiasm of two researchers, Jelle Zerlinga de Boer and Johan Varekamp of Wesleyan University and I will forever be grateful to them for sharing their enthusiasm with me for Volcán Copahue and its environs. Sadly, as this book was going to press, I learned of Jelle's passing. He was the person who first sparked my interest in this tree. He will be greatly missed.

I enternally thank Ernesto Maletti for his generosity and camaraderie and for sharing his love and knowledge of the pehuén. As is evident in these pages, his help was infinite. My thanks also to Rubén Vargas for his adventurous spirit and guidance throughout the steep terrain of the Andes.

Charlotte Currier has my appreciation for the time and insight she gave in helping me to coherently bring these words and ideas together and similarly, Anna Pillow and Davi Leventhal for their creative contributions in bringing this book to fruition.

I must certainly acknowledge the Vermont Studio Center and the Sacatar Foundation not just for the space and time to write which they have granted me over the years but for the spectacular landscapes they provide. These environs have not only nurtured my work but also my soul. The same goes for the Andrews Forest Writers' Residency, which is located in a place where the trees are also very old and very tall. There is no quieter place to write than in a forest.

This book is dedicated to the memory of Tina Song, a dear friend and steadfast supporter and to the memory of Ginger Roberts; two dear friends who both loved to read as much as they loved to travel and who left far too soon.

Lightning Source UK Ltd.
Milton Keynes UK
UKOW07f0201031216
288821UK00020B/133/P